Electricity and Magnetism for Mathematicians

This text is an introduction to some of the mathematical wonders of Maxwell's equations. These equations led to the prediction of radio waves, the realization that light is a type of electromagnetic wave, and the discovery of the special theory of relativity. In fact, almost all current descriptions of the fundamental laws of the universe can be viewed as deep generalizations of Maxwell's equations. Even more surprising is that these equations and their generalizations have led to some of the most important mathematical discoveries of the past thirty years. It seems that the mathematics behind Maxwell's equations is endless.

The goal of this book is to explain to mathematicians the underlying physics behind electricity and magnetism and to show their connections to mathematics. Starting with Maxwell's equations, the reader is led to such topics as the special theory of relativity, differential forms, quantum mechanics, manifolds, tangent bundles, connections, and curvature.

THOMAS A. GARRITY is the William R. Kenan, Jr. Professor of Mathematics at Williams, where he was the director of the Williams Project for Effective Teaching for many years. In addition to a number of research papers, he has authored or coauthored two other books, *All the Mathematics You Missed [But Need to Know for Graduate School]* and *Algebraic Geometry: A Problem Solving Approach*. Among his awards and honors is the MAA Deborah and Franklin Tepper Haimo Award for outstanding college or university teaching.

ELECTRICITY AND MAGNETISM
FOR MATHEMATICIANS

A Guided Path from Maxwell's
Equations to Yang-Mills

THOMAS A. GARRITY

Williams College, Williamstown, Massachusetts
with illustrations by Nicholas Neumann-Chun

CAMBRIDGE
UNIVERSITY PRESS

CAMBRIDGE
UNIVERSITY PRESS

University Printing House, Cambridge CB2 8BS, United Kingdom
One Liberty Plaza, 20th Floor, New York, NY 10006, USA
477 Williamstown Road, Port Melbourne, VIC 3207, Australia
4843/24, 2nd Floor, Ansari Road, Daryaganj, Delhi – 110002, India
79 Anson Road, #06-04/06, Singapore 079906

Cambridge University Press is part of the University of Cambridge.

It furthers the University's mission by disseminating knowledge in the pursuit of
education, learning, and research at the highest international levels of excellence.

www.cambridge.org
Information on this title: www.cambridge.org/9781107435162

First published 2015

A catalog record for this publication is available from the British Library.

Library of Congress Cataloging in Publication Data
Garrity, Thomas A., 1959– author.
Electricity and magnetism for mathematicians : a guided path from Maxwell's
equations to Yang-Mills / Thomas A. Garrity, Williams College, Williamstown,
Massachusetts; with illustrations by Nicholas Neumann-Chun.
pages   cm
Includes bibliographical references and index.
ISBN 978-1-107-07820-8 (hardback) – ISBN 978-1-107-43516-2 (paperback)
1. Electromagnetic theory–Mathematics–Textbooks.   I. Title.
QC670.G376   2015
537.01′51–dc23   2014035298

ISBN 978-1-107-07820-8 Hardback
ISBN 978-1-107-43516-2 Paperback

Contents

List of Symbols   *page* xi
Acknowledgments   xiii

1   A Brief History   **1**
   1.1   Pre-1820: The Two Subjects of Electricity and Magnetism   1
   1.2   1820–1861: The Experimental Glory Days of
           Electricity and Magnetism   2
   1.3   Maxwell and His Four Equations   2
   1.4   Einstein and the Special Theory of Relativity   2
   1.5   Quantum Mechanics and Photons   3
   1.6   Gauge Theories for Physicists:
           The Standard Model   4
   1.7   Four-Manifolds   5
   1.8   This Book   7
   1.9   Some Sources   7

2   Maxwell's Equations   **9**
   2.1   A Statement of Maxwell's Equations   9
   2.2   Other Versions of Maxwell's Equations   12
       2.2.1   Some Background in Nabla   12
       2.2.2   Nabla and Maxwell   14
   2.3   Exercises   14

3   Electromagnetic Waves   **17**
   3.1   The Wave Equation   17
   3.2   Electromagnetic Waves   20
   3.3   The Speed of Electromagnetic Waves Is Constant   21
       3.3.1   Intuitive Meaning   21

	3.3.2	Changing Coordinates for the Wave Equation	22
3.4	Exercises		25

4 Special Relativity   **27**
4.1	Special Theory of Relativity		27
4.2	Clocks and Rulers		28
4.3	Galilean Transformations		31
4.4	Lorentz Transformations		32
	4.4.1	A Heuristic Approach	32
	4.4.2	Lorentz Contractions and Time Dilations	35
	4.4.3	Proper Time	36
	4.4.4	The Special Relativity Invariant	37
	4.4.5	Lorentz Transformations, the Minkowski Metric, and Relativistic Displacement	38
4.5	Velocity and Lorentz Transformations		43
4.6	Acceleration and Lorentz Transformations		45
4.7	Relativistic Momentum		46
4.8	Appendix: Relativistic Mass		48
	4.8.1	Mass and Lorentz Transformations	48
	4.8.2	More General Changes in Mass	51
4.9	Exercises		52

5 Mechanics and Maxwell's Equations   **56**
5.1	Newton's Three Laws		56
5.2	Forces for Electricity and Magnetism		58
	5.2.1	$F = q(E + v \times B)$	58
	5.2.2	Coulomb's Law	59
5.3	Force and Special Relativity		60
	5.3.1	The Special Relativistic Force	60
	5.3.2	Force and Lorentz Transformations	61
5.4	Coulomb + Special Relativity + Charge Conservation = Magnetism		62
5.5	Exercises		65

6 Mechanics, Lagrangians, and the Calculus of Variations   **70**
6.1	Overview of Lagrangians and Mechanics		70
6.2	Calculus of Variations		71
	6.2.1	Basic Framework	71
	6.2.2	Euler-Lagrange Equations	73
	6.2.3	More Generalized Calculus of Variations Problems	77
6.3	A Lagrangian Approach to Newtonian Mechanics		78

6.4	Conservation of Energy from Lagrangians	83
6.5	Noether's Theorem and Conservation Laws	85
6.6	Exercises	86
7	**Potentials**	**88**
7.1	Using Potentials to Create Solutions for Maxwell's Equations	88
7.2	Existence of Potentials	89
7.3	Ambiguity in the Potential	91
7.4	Appendix: Some Vector Calculus	91
7.5	Exercises	95
8	**Lagrangians and Electromagnetic Forces**	**98**
8.1	Desired Properties for the Electromagnetic Lagrangian	98
8.2	The Electromagnetic Lagrangian	99
8.3	Exercises	101
9	**Differential Forms**	**103**
9.1	The Vector Spaces $\Lambda^k(\mathbb{R}^n)$	103
9.1.1	A First Pass at the Definition	103
9.1.2	Functions as Coefficients	106
9.1.3	The Exterior Derivative	106
9.2	Tools for Measuring	109
9.2.1	Curves in $\mathbb{R}^3$	109
9.2.2	Surfaces in $\mathbb{R}^3$	111
9.2.3	k-manifolds in $\mathbb{R}^n$	113
9.3	Exercises	115
10	**The Hodge $\star$ Operator**	**119**
10.1	The Exterior Algebra and the $\star$ Operator	119
10.2	Vector Fields and Differential Forms	121
10.3	The $\star$ Operator and Inner Products	122
10.4	Inner Products on $\Lambda(\mathbb{R}^n)$	123
10.5	The $\star$ Operator with the Minkowski Metric	125
10.6	Exercises	127
11	**The Electromagnetic Two-Form**	**130**
11.1	The Electromagnetic Two-Form	130
11.2	Maxwell's Equations via Forms	130
11.3	Potentials	131
11.4	Maxwell's Equations via Lagrangians	132
11.5	Euler-Lagrange Equations for the Electromagnetic Lagrangian	136
11.6	Exercises	139

12 Some Mathematics Needed for Quantum Mechanics   **142**
   12.1   Hilbert Spaces   142
   12.2   Hermitian Operators   149
   12.3   The Schwartz Space   153
       12.3.1   The Definition   153
       12.3.2   The Operators $q(f) = xf$ and $p(f) = -i\,\mathrm{d}f/\mathrm{d}x$ 155
       12.3.3 $\mathcal{S}(\mathbb{R})$ Is Not a Hilbert Space   157
   12.4   Caveats: On Lebesgue Measure, Types of Convergence,
        and Different Bases   159
   12.5   Exercises   160

13 Some Quantum Mechanical Thinking   **163**
   13.1   The Photoelectric Effect: Light as Photons   163
   13.2   Some Rules for Quantum Mechanics   164
   13.3   Quantization   170
   13.4   Warnings of Subtleties   172
   13.5   Exercises   172

14 Quantum Mechanics of Harmonic Oscillators   **176**
   14.1   The Classical Harmonic Oscillator   176
   14.2   The Quantum Harmonic Oscillator   179
   14.3   Exercises   184

15 Quantizing Maxwell's Equations   **186**
   15.1   Our Approach   186
   15.2   The Coulomb Gauge   187
   15.3   The "Hidden" Harmonic Oscillator   193
   15.4   Quantization of Maxwell's Equations   195
   15.5   Exercises   197

16 Manifolds   **201**
   16.1   Introduction to Manifolds   201
       16.1.1   Force = Curvature   201
       16.1.2   Intuitions behind Manifolds   201
   16.2   Manifolds Embedded in $\mathbb{R}^n$ 203
       16.2.1   Parametric Manifolds   203
       16.2.2   Implicitly Defined Manifolds   205
   16.3   Abstract Manifolds   206
       16.3.1   Definition   206
       16.3.2   Functions on a Manifold   212
   16.4   Exercises   212

17 Vector Bundles   **214**
  17.1 Intuitions   214
  17.2 Technical Definitions   216
         17.2.1 The Vector Space $\mathbb{R}^k$ 216
         17.2.2 Definition of a Vector Bundle   216
  17.3 Principal Bundles   219
  17.4 Cylinders and Möbius Strips   220
  17.5 Tangent Bundles   222
         17.5.1 Intuitions   222
         17.5.2 Tangent Bundles for Parametrically Defined
                 Manifolds   224
         17.5.3 $T(\mathbb{R}^2)$ as Partial Derivatives   225
         17.5.4 Tangent Space at a Point of an Abstract Manifold 227
         17.5.5 Tangent Bundles for Abstract Manifolds   228
  17.6 Exercises   230

18 Connections   **232**
  18.1 Intuitions   232
  18.2 Technical Definitions   233
         18.2.1 Operator Approach   233
         18.2.2 Connections for Trivial Bundles   237
  18.3 Covariant Derivatives of Sections   240
  18.4 Parallel Transport: Why Connections Are Called
        Connections   243
  18.5 Appendix: Tensor Products of Vector Spaces   247
         18.5.1 A Concrete Description   247
         18.5.2 Alternating Forms as Tensors   248
         18.5.3 Homogeneous Polynomials as Symmetric Tensors   250
         18.5.4 Tensors as Linearizations of Bilinear Maps   251
  18.6 Exercises   253

19 Curvature   **257**
  19.1 Motivation   257
  19.2 Curvature and the Curvature Matrix   258
  19.3 Deriving the Curvature Matrix   260
  19.4 Exercises   261

20 Maxwell via Connections and Curvature   **263**
  20.1 Maxwell in Some of Its Guises   263
  20.2 Maxwell for Connections and Curvature   264
  20.3 Exercises   266

21   The Lagrangian Machine, Yang-Mills, and Other Forces   267
        21.1   The Lagrangian Machine   267
        21.2   U(1) Bundles   268
        21.3   Other Forces   269
        21.4   A Dictionary   270
        21.5   Yang-Mills Equations   272

Bibliography   275
Index   279

Color plates follow page 234

List of Symbols

Symbol	Name
∇	nabla
$\triangle$	Laplacian
T	transpose
$\in$	element of
$O(3, \mathbb{R})$	orthogonal group
$\mathbb{R}$	real numbers
$\rho(\cdot, \cdot)$	Minkowski metric
$\wedge^k(\mathbb{R}^n)$	k-forms on $\mathbb{R}^n$
$\wedge$	wedge
$\circ$	composed with
$\star$	star operator
$\mathcal{H}$	Hilbert space
$\langle \cdot, \cdot \rangle$	inner product
$\mathbb{C}$	complex numbers
$L^2[0, 1]$	square integrable functions
$*$	adjoint
$\subset$	subset of
$\mathcal{S}$	Schwartz space
h	Planck constant
$\cap$	set intersection
$\cup$	set union
$GL(k, \mathbb{R})$	general linear group
C_p^∞	germ of the sheaf of differentiable functions
$\Gamma(E)$	space of all sections of E
∇	connection
$\otimes$	tensor product
$\odot$	symmetric tensor product

Acknowledgments

There are many people who have helped in the preparing of this book. First off, an earlier draft was used as the text for a course at Williams College in the fall of 2009. In this class, Ben Atkinson, Ran Bi, Victoria Borish, Aaron Ford, Sarah Ginsberg, Charlotte Healy, Ana Inoa, Stephanie Jensen, Dan Keneflick, Murat Kologlu, Edgar Kosgey, Jackson Lu, Makisha Maier, Alex Massicotte, Merideth McClatchy, Nicholas Neumann-Chun, Ellen Ramsey, Margaret Robinson, Takuta Sato, Anders Schneider, Meghan Shea, Joshua Solis, Elly Tietsworth, Stephen Webster, and Qiao Zhang provided a lot of feedback. In particular Stephen Webster went through the entire manuscript again over the winter break of 2009–2010. I would like to thank Weng-Him Cheung, who went through the whole manuscript in the fall of 2013. I would also like to thank Julia Cline, Michael Mayer, Cesar Melendez, and Emily Wickstrom, all of whom took a course based on this text at Williams in the fall of 2013, for helpful comments.

Anyone who would like to teach a course based on this text, please let me know (tgarrity@williams.edu). In particular, there are write-ups of the solutions for many of the problems. I have used the text for three classes, so far. The first time the prerequisites were linear algebra and multivariable calculus. For the other classes, the perquisites included real analysis. The next time I teach this course, I will return to only requiring linear algebra and multivariable calculus. As Williams has fairly short semesters (about twelve to thirteen weeks), we covered only the first fifteen chapters, with a brief, rapid-fire overview of the remaining topics.

In the summer of 2010, Nicholas Neumann-Chun proofread the entire manuscript, created its diagrams, and worked a lot of the homework problems. He gave many excellent suggestions.

My Williams colleague Steven Miller also carefully read a draft, helping tremendously. Also from Williams, Lori Pedersen went through the text a few times and provided a lot of solutions of the homework problems. Both William Wootters and David Tucker-Smith, from the Williams Physics Department, also gave a close reading of the manuscript; both provided key suggestions for improving the physics in the text.

Robert Kotiuga helped with the general exposition and especially in giving advice on the history of the subject.

I would like to thank Gary Knapp, who not only went through the whole text, providing excellent feedback, but who also suggested a version of the title. Both Dakota Garrity and Logan Garrity caught many errors and typos in the final draft. Each also gave excellent suggestions for improving the exposition.

I also would like to thank my editor, Lauren Cowles, who has provided support through this whole project.

The referees also gave much-needed advice.

I am grateful for all of their help.

1

A Brief History

Summary: The unification of electricity, magnetism, and light by James Maxwell in the 1800s was a landmark in human history and has continued even today to influence technology, physics, and mathematics in profound and surprising ways. Its history (of which we give a brief overview in this chapter) has been and continues to be studied by historians of science.

1.1. Pre-1820: The Two Subjects of Electricity and Magnetism

Who knows when our ancestors first became aware of electricity and magnetism? I imagine a primitive cave person, wrapped up in mastodon fur, desperately trying to stay warm in the dead of winter, suddenly seeing a spark of static electricity. Maybe at the same time in our prehistory someone felt a small piece of iron jump out of their hand toward a lodestone. Certainly lightning must have been inspiring and frightening, as it still is.

But only recently (meaning in the last four hundred years) have these phenomena been at all understood. Around 1600, William Gilbert wrote his infuential *De Magnete*, in which he argued that the earth was just one big magnet. In the mid-1700s, Benjamin Franklin showed that lightning was indeed electricity. Also in the 1700s Coulomb's law was discovered, which states that the force F between two stationary charges is

$$F = \frac{q_1 q_2}{r^2},$$

where q_1 and q_2 are the charges and r is the distance between the charges (after choosing correct units). Further, in the 1740s, Leyden jars were invented to store electric charge. Finally, still in the 1700s, Galvani and Volta, independently, discovered how to generate electric charges, with the invention of galvanic, or voltaic, cells (batteries).

1.2. 1820–1861: The Experimental Glory Days of
Electricity and Magnetism

In 1820, possibly during a lecture, Hans Christian Oersted happened to move a compass near a wire that carried a current. He noticed that the compass's needle jumped. People knew that compasses worked via magnetism and at the same time realized that current was flowing electricity. Oersted found solid proof that electricity and magnetism were linked.

For the next forty or so years amazing progress was made finding out how these two forces were related. Most of this work was rooted in experiment. While many scientists threw themselves into this hunt, Faraday stands out as a truly profound experimental scientist. By the end of this era, most of the basic empirical connections between electricity and magnetism had been discovered.

1.3. Maxwell and His Four Equations

In the early 1860s, James Clerk Maxwell wrote down his four equations that linked the electric field with the magnetic field. (The real history is quite a bit more complicated.) These equations contain within them the prediction that there are electromagnetic waves, traveling at some speed c. Maxwell observed that this speed c was close to the observed speed of light. This led him to make the spectacular conjecture that light is an electromagnetic wave. Suddenly, light, electricity, and magnetism were all part of the same fundamental phenomenon.

Within twenty years, Hertz had experimentally shown that light was indeed an electromagnetic wave. (As seen in Chapter 6 of [27], the actual history is not quite such a clean story.)

1.4. Einstein and the Special Theory of Relativity

All electromagnetic waves, which after Maxwell were known to include light waves, have a remarkable yet disturbing property: These waves travel at a fixed speed c. This fact was not controversial at all, until it was realized that this speed was independent of any frame of reference.

To make this surprise more concrete, we turn to Einstein's example of shining lights on trains. (No doubt today the example would be framed in terms of airplanes or rocket ships.) Imagine you are on a train traveling at 60 miles per hour. You turn on a flashlight and point it in the same direction as the train is moving. To you, the light moves at a speed of c (you think your speed is zero miles per hour). To someone on the side of the road, the light should move at a speed of 60 miles per hour $+c$. But according to Maxwell's equations, it does not. The observer off the train will actually see the light move at the same speed c, which is no different from your observation on the train. This is wacky and suggests that Maxwell's equations must be wrong.

In actual experiments, though, it is our common sense (codified in Newtonian mechanics) that is wrong. This led Albert Einstein, in 1905, to propose an entirely new theory of mechanics, the special theory. In large part, Einstein discovered the special theory because he took Maxwell's equations seriously as a statement about the fundamental nature of reality.

1.5. Quantum Mechanics and Photons

What is light? For many years scientists debated whether light was made up of particles or of waves. After Maxwell (and especially after Hertz's experiments showing that light is indeed a type of electromagnetic wave), it seemed that the debate had been settled. But in the late nineteenth century, a weird new phenomenon was observed. When light was shone on certain metals, electrons were ejected from the metal. Something in light carried enough energy to forcibly eject electrons from the metal. This phenomenon is called the *photoelectric effect*. This alone is not shocking, as it was well known that traditional waves carried energy. (Many of us have been knocked over by ocean waves at the beach.) In classical physics, though, the energy carried by a traditional wave is proportional to the wave's amplitude (how high it gets). But in the photoelectric effect, the energy of the ejected electrons is proportional not to the amplitude of the light wave but instead to the light's frequency. This is a decidedly non-classical effect, jeopardizing a wave interpretation for light.

In 1905, in the same year that he developed the Special Theory of Relativity, Einstein gave an interpretation to light that seemed to explain the photoelectric effect. Instead of thinking of light as a wave (in which case, the energy would have to be proportional to the light's amplitude), Einstein assumed that light is made of particles, each of which has energy proportional to the frequency, and showed that this assumption leads to the correct experimental predictions.

In the context of other seemingly strange experimental results, people started to investigate what is now called quantum mechanics, amassing a number of partial explanations. Suddenly, over the course of a few years in the mid-1920s, Born, Dirac, Heisenberg, Jordan, Schrödinger, von Neumann, and others worked out the complete theory, finishing the first quantum revolution. We will see that this theory indeed leads to the prediction that light must have properties of both waves and particles.

1.6. Gauge Theories for Physicists:
The Standard Model

At the end of the 1920s, gravity and electromagnetism were the only two known forces. By the end of the 1930s, both the strong force and the weak force had been discovered.

In the nucleus of an atom, protons and neutrons are crammed together. All of the protons have positive charge. The rules of electromagnetism would predict that these protons would want to explode away from each other, but this does not happen. It is the strong force that holds the protons and neutrons together in the nucleus, and it is called such since it must be strong enough to overcome the repelling force of electromagnetism.

The weak force can be seen in the decay of the neutron. If a neutron is just sitting around, after ten or fifteen minutes it will decay into a proton, an electron, and another elementary particle (the electron anti-neutrino, to be precise). This could not be explained by the other forces, leading to the discovery of this new force.

Since both of these forces were basically described in the 1930s, their theories were quantum mechanical. But in the 1960s, a common framework for the weak force and the electromagnetic force was worked out (resulting in Nobel Prizes for Abdus Salam, Sheldon Glashow, and Steven Weinberg in 1979). In fact, this framework can be extended to include the strong force. This common framework goes by the name of the *standard model*. (It does not include gravity.)

Much earlier, in the 1920s, the mathematician Herman Weyl attempted to unite gravity and electromagnetism, by developing what he called a *gauge theory*.   While it quickly was shown not to be physically realistic, the underlying idea was sufficiently intriguing that it resurfaced in the early 1950s in the work of Yang and Mills, who were studying the strong force. The underlying mathematics of their work is what led to the unified electro-weak force and the standard model.

Weyl's gauge theory was motivated by symmetry. He used the word "gauge" to suggest different gauges for railroad tracks.   His work was motivated by the desire to shift from global symmetries to local symmetries. We will start with global symmetries. Think of the room you are sitting in. Choose a corner and label this the origin. Assume one of the edges is the x-axis, another the y-axis, and the third the z-axis. Put some unit of length on these edges. You can now uniquely label any point in the room by three coordinate values.

Of course, someone else might have chosen a different corner as the origin, different coordinate axes, or different units of length. In fact, any point in the room (or, for that matter, any point in space) could be used as the origin, and so on. There are an amazing number of different choices.

Now imagine a bird flying in the room. With your coordinate system, you could describe the path of the bird's flight by a curve $(x(t), y(t), z(t))$. Someone else, with a different coordinate system, will describe the flight of the bird by three totally different functions. The flight is the same (after all, the bird does not care what coordinate system you are using), but the description is different. By changing coordinates, we can translate from one coordinate system into the other. This is a global change of coordinates. Part of the deep insight of the theory of relativity, as we will see, is that which coordinate changes are allowed has profound effects on the description of reality.

Weyl took this one step further. Instead of choosing one global coordinate system, he proposed that we could choose different coordinate systems at each point of space but that all of these local coordinate systems must be capable of being suitably patched together. Weyl called this patching "choosing a gauge."

1.7. Four-Manifolds

During the 1950s, 1960s, and early 1970s, when physicists were developing what they called gauge theory, leading to the standard model, mathematicians were developing the foundations of differential geometry. (Actually this work on differential geometry went back quite a bit further than the 1950s.) This mainly involved understanding the correct nature of curvature, which, in turn, as we will see, involves understanding the nature of connections. But sometime in the 1960s or 1970s, people must have begun to notice uncanny similarities between the physicists' gauges and the mathematicians' connections. Finally, in 1975, Wu and Yang [69] wrote out the dictionary between the two languages (this is the same Yang who was part of Yang-Mills). This alone was amazing. Here the foundations of much of modern physics were shown to be the same as the foundations of much of differential geometry.

Through most of the twentieth century, when math and physics interacted, overwhelmingly it was the case that math shaped physics:

$$\text{Mathematics} \Rightarrow \text{Physics}$$

Come the early 1980s, the arrow was reversed. Among all possible gauges, physicists pick out those that are Yang-Mills, which are in turn deep generalizations of Maxwell's equations. By the preceding dictionary, connections that satisfy Yang-Mills should be special.

This leads us to the revolutionary work of Simon Donaldson. He was interested in four-dimensional manifolds. On a four-manifold, there is the space of all possible connections. (We are ignoring some significant facts.) This space is infinite dimensional and has little structure. But then Donaldson decided to take physicists seriously. He looked at those connections that were Yang-Mills. (Another common term used is "instantons.") At the time, there was no compelling mathematical reason to do this. Also, his four-manifolds were not physical objects and had no apparent link with physics. Still, he looked at Yang-Mills connections and discovered amazing, deeply surprising structure, such as that these special Yang-Mills connections form a five-dimensional space, which has the original four-manifold as part of its boundary. (Here we are coming close to almost criminal simplification, but the underlying idea that the Yang-Mills connections are linked to a five-manifold that has the four-manifold as part of its boundary is correct.) This work shocked much of the mathematical world and transformed four-manifold theory from a perfectly respectable area of mathematics into one of its hottest branches. In awarding Donaldson a Field's Medal in 1986, Atiyah [1] wrote:

> The surprise produced by Donaldson's result was accentuated by the fact that his methods were completely new and were borrowed from theoretical physics, in the form of Yang-Mills equations. ... Several mathematicians (including myself) worked on instantons and felt very pleased that they were able to assist physics in this way. Donaldson, on the other hand, conceived the daring idea of reversing this process and of using instantons on a general 4-manifold as a new geometrical tool.

Many of the finest mathematicians of the 1980s started working on developing this theory, people such as Atiyah, Bott, Uhlenbeck, Taubes, Yau, Kobayashi, and others.

Not only did this work produce some beautiful mathematics, it changed how math could be done. Now we have

$$\text{Physics} \Rightarrow \text{Mathematics}$$

an approach that should be called *physical mathematics* (a term first coined by Kishore Marathe, according to [70]: This text by Zeidler is an excellent place to begin to see the power behind the idea of physical mathematics).

Physical mathematics involves taking some part of the real world that is physically important (such as Maxwell's equations), identifying the underlying mathematics, and then taking that mathematics seriously, even in contexts far removed from the natural world. This has been a major theme of mathematics since the 1980s, led primarily by the brilliant work of Edward Witten. When Witten won his Field's Medal in 1990, Atiyah [2] wrote:

> Although (Witten) is definitely a physicist his command of mathematics is rivaled by few mathematicians, and his ability to interpret physical ideas in mathematical form is quite unique. Time and again he has surprised the mathematical community by a brilliant application of physical insight leading to new and deep mathematical theorems.

The punchline is that mathematicians should take seriously underlying mathematical structure of the real world, even in non-real world situations. In essence, nature is a superb mathematician.

1.8. This Book

There is a problem with this revolution of physical mathematics. How can any mere mortal master both physics and mathematics? The answer, of course, is you cannot. This book is a compromise. We concentrate on the key underlying mathematical concepts behind the physics, trying at the same time to explain just enough of the real world to justify the use of the mathematics. By the end of this book, I hope the reader will be able to start understanding the work needed to understand Yang-Mills.

1.9. Some Sources

One way to learn a subject is to study its history. That is not the approach we are taking. There are a number of good, accessible books, though. Stephen J. Blundell's *Magnetism: A Very Short Introduction* [4] is excellent for a popular general overview. For more technical histories of the early days of electromagnetism, I would recommend Steinle's article "Electromagnetism and Field Physics" [62] and Buchwald's article "Electrodynamics from Thomson and Maxwell to Hertz" [7].

Later in his career, Abraham Pais wrote three excellent books covering much of the history of twentieth century physics. His *Subtle Is the Lord: The Science and the Life of Albert Einstein* [51] is a beautiful scientific biography of Einstein, which means that it is also a history of much of what was important in physics in the first third of the 1900s. His *Niels Bohr's Times: In Physics, Philosophy, and Polity* [52] is a scientific biography of Bohr, and hence a good overview of the history of early quantum mechanics. His *Inward Bound* [53] is a further good reference for the development of quantum theory and particle physics.

It appears that the ideas of special relativity were "in the air" around 1905. For some of the original papers by Einstein, Lorentz, Minkowski, and Weyl, there is the collection [19]. Poincaré was also actively involved in the early days of special relativity. Recently two biographies of Poincaré have been written: Gray's *Henri Poincaré: A Scientific Biography* [27] and Verhulst's *Henri Poincaré: Impatient Genius* [67]. There is also the still interesting paper of Poincaré that he gave at the World's Fair in Saint Louis in 1904, which has recently been reprinted [54].

At the end of this book, we reach the beginnings of gauge theory. In [50], O'Raifeartaigh has collected some of the seminal papers in the development of gauge theory. We encourage the reader to look at the web page of Edward Witten for inspiration. I would also encourage people to look at many of the expository papers on the relationship between mathematics and physics in volume 6 of the collected works of Atiyah [3] and at those in volume 4 of the collected works of Bott [5]. (In fact, perusing all six volumes of Atiyah's collected works and all four volumes of Bott's is an excellent way to be exposed to many of the main themes of mathematics of the last half of the twentieth century.) Finally, there is the wonderful best seller *The Elegant Universe* by Brian Greene [28].

2

Maxwell's Equations

Summary: The primary goal of this chapter is to state Maxwell's equations. We will then see some of their implications, which will allow us to give alternative descriptions for Maxwell's equations, providing us in turn with a review of some of the basic formulas in multivariable calculus.

2.1. A Statement of Maxwell's Equations

Maxwell's equations link together three vector fields and a real-valued function. Let

$$E = E(x,y,z,t) = (E_1(x,y,z,t), E_2(x,y,z,t), E_3(x,y,z,t))$$

and

$$B = B(x,y,z,t) = (B_1(x,y,z,t), B_2(x,y,z,t), B_3(x,y,z,t))$$

be two vector fields with spacial coordinates (x,y,z) and time coordinate t. Here E represents the electric field while B represents the magnetic field. The third vector field is

$$j(x,y,z,t) = (j_1(x,y,z,t), j_2(x,y,z,t), j_3(x,y,z,t)),$$

which represents the current (the direction and the magnitude of the flow of electric charge). Finally, let

$$\rho(x,y,z,t)$$

be a function representing the charge density. Let c be a constant. (Here c is the speed of light in a vacuum.) Then these three vector fields and this function

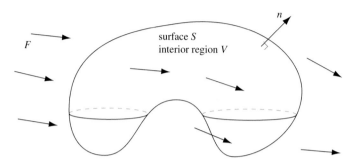

Figure 2.1

satisfy Maxwell's equations if

$$\mathrm{div}(E) = \rho$$

$$\mathrm{curl}(E) = -\frac{\partial B}{\partial t}$$

$$\mathrm{div}(B) = 0$$

$$c^2\,\mathrm{curl}(B) = j + \frac{\partial E}{\partial t}.$$

(Review of the curl, the divergence, and other formulas from multivariable calculus is in the next section.)

We can reinterpret these equations in terms of integrals via various Stokes-type theorems. For example, if V is a compact region in space with smooth boundary surface S, as in Figure 2.1, then for any vector field F we know from the Divergence Theorem that

$$\int\!\!\int_S F \cdot n\, \mathrm{d}A = \int\!\!\int\!\!\int_V \mathrm{div}(F)\, \mathrm{d}x\mathrm{d}y\mathrm{d}z,$$

where n is the unit outward normal of the surface S.

In words, this theorem says that the divergence of a vector field measures how much of the field is flowing out of a region.

Then the first of Maxwell's equations can be restated as

$$\int\!\!\int_S E \cdot n\, \mathrm{d}A = \int\!\!\int\!\!\int \mathrm{div}(E)\, \mathrm{d}x\mathrm{d}y\mathrm{d}z$$

$$= \int\!\!\int\!\!\int \rho(x,y,z,t)\, \mathrm{d}x\mathrm{d}y\mathrm{d}z$$

$$= \text{total charge inside the region } V.$$

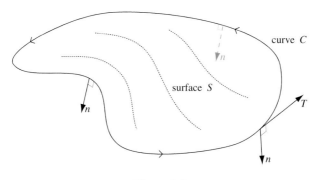

Figure 2.2

Likewise, the third of Maxwell's equations is:

$$\int\int_S B \cdot n \, dA = \int\int\int \text{div}(B) \, dx\,dy\,dz$$

$$= \int\int\int 0 \, dx\,dy\,dz$$

$$= 0$$

$$= \text{There is no magnetic charge inside the region } V.$$

This is frequently stated as "There are no magnetic monopoles," meaning there is no real physical notion of magnetic density.

The second and fourth of Maxwell's equations have similar integral interpretations. Let C be a smooth curve in space that is the boundary of a smooth surface S, as in Figure 2.2. Let T be a unit tangent vector of C. Choose a normal vector field n for S so that the cross product $T \times n$ points into the surface S.

Then the classical Stokes Theorem states that for any vector field F, we have

$$\int_C F \cdot T \, ds = \int\int_S \text{curl}(F) \cdot n \, dA.$$

This justifies the intuition that the curl of a vector field measures how much the vector field F wants to twirl.

Then the second of Maxwell's equations is equivalent to

$$\int_C E \cdot T \, ds = -\int\int_S \frac{\partial B}{\partial t} \cdot n \, dA.$$

Thus the magnetic field B is changing in time if and only if the electric field E is curling.

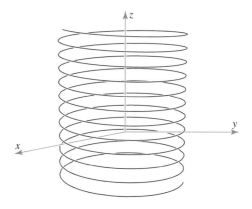

Figure 2.3

This is the mathematics underlying how to create current in a wire by moving a magnet. Consider a coil of wire, centered along the z-axis (i.e., along the vector $k = (0,0,1)$).

The wire is coiled (almost) in the xy-plane. Move a magnet through the middle of this coil. This means that the magnetic field B is changing in time in the direction k. Thanks to Maxwell, this means that the curl of the electric field E will be non-zero and will point in the direction k. But this means that the actual vector field E will be "twirling" in the xy-plane, making the electrons in the coil move, creating a current.

This is in essence how a hydroelectric dam works. Water from a river is used to move a magnet through a coil of wire, creating a current and eventually lighting some light bulb in a city far away.

The fourth Maxwell equation gives

$$c^2 \int_C B \cdot T \, ds = \int \int_S \left(j + \frac{\partial E}{\partial t} \right) \cdot n \, dA.$$

Here current and a changing electric field are linked to the curl of the magnetic field.

2.2. Other Versions of Maxwell's Equations

2.2.1. Some Background in Nabla

This section is meant to be both a review and a listing of some of the standard notations that people use. The symbol ∇ is pronounced "nabla" (sometimes ∇

is called "del"). Let

$$\nabla = \left(\frac{\partial}{\partial x}, \frac{\partial}{\partial y}, \frac{\partial}{\partial z}\right)$$
$$= i\frac{\partial}{\partial x} + j\frac{\partial}{\partial y} + k\frac{\partial}{\partial z}$$

where $i = (1,0,0)$, $j = (0,1,0)$, and $k = (0,0,1)$. Then for any function $f(x,y,z)$, we set the gradient to be

$$\nabla f = \left(\frac{\partial f}{\partial x}, \frac{\partial f}{\partial y}, \frac{\partial f}{\partial z}\right)$$
$$= i\frac{\partial f}{\partial x} + j\frac{\partial f}{\partial y} + k\frac{\partial f}{\partial z}.$$

For a vector field

$$F = F(x,y,z) = (F_1(x,y,z), F_2(x,y,z), F_3(x,y,z))$$
$$= (F_1, F_2, F_3)$$
$$= F_1 \cdot i + F_2 \cdot j + F_3 \cdot k,$$

define the divergence to be:

$$\nabla \cdot F = \text{div}(F)$$
$$= \frac{\partial F_1}{\partial x} + \frac{\partial F_2}{\partial y} + \frac{\partial F_3}{\partial z}.$$

The curl of a vector field in this notation is

$$\nabla \times F = \text{curl}(F)$$
$$= \det \begin{pmatrix} i & j & k \\ \frac{\partial}{\partial x} & \frac{\partial}{\partial y} & \frac{\partial}{\partial z} \\ F_1 & F_2 & F_3 \end{pmatrix}$$
$$= \left(\frac{\partial F_3}{\partial y} - \frac{\partial F_2}{\partial z}, -\left(\frac{\partial F_3}{\partial x} - \frac{\partial F_1}{\partial z}\right), \frac{\partial F_2}{\partial x} - \frac{\partial F_1}{\partial y}\right).$$

2.2.2. Nabla and Maxwell

Using the nabla notation, Maxwell's equations have the form

$$\nabla \cdot E = \rho$$

$$\nabla \times E = -\frac{\partial B}{\partial t}$$

$$\nabla \cdot B = 0$$

$$c^2 \nabla \times B = j + \frac{\partial E}{\partial t}$$

Though these look like four equations, when written out they actually form eight equations. This is one of the exercises in the following section.

2.3. Exercises

The first few problems are exercises in the nabla machinery and in the basics of vector fields.

Exercise 2.3.1. *For the function* $f(x,y,z) = x^2 + y^3 + xy^2 + 4z$, *compute* grad(f), *which is the same as computing* $\nabla(f)$.

Exercise 2.3.2. *a. Sketch, in the xy-plane, some representative vectors making up the vector field*

$$F(x,y,z) = (F_1, F_2, F_3) = (x,y,z),$$

at the points

$$(1,0,0),(1,1,0),(0,1,0),(-1,1,0),(-1,0,0),(-1,-1,0),(0,-1,0),(1,-1,0).$$

   b. Find div(F) $= \nabla \cdot F$.
   c. Find curl(F) $= \nabla \times F$.

Comment: Geometrically the vector field $F(x,y,z) = (x,y,z)$ is spreading out but not "twirling" or "curling" at all, as is reflected in the calculations of its divergence and curl.

Exercise 2.3.3. *a. Sketch, in the xy-plane, some representative vectors making up the vector field*

$$F(x,y,z) = (F_1, F_2, F_3) = (-y, x, 0),$$

at the points

$$(1,0,0),(1,1,0),(0,1,0),(-1,1,0),(-1,0,0),(-1,-1,0),(0,-1,0),(1,-1,0).$$

b. *Find* div(F) = ∇ · F.
c. *Find* curl(F) = ∇ × F.

Comment: As compared to the vector field in the previous exercise, this vector field $F(x,y,z) = (-y,x,0)$ is not spreading out at all but does "twirl" in the xy-plane. Again, this is reflected in the divergence and curl.

Exercise 2.3.4. *Write out Maxwell's equations in local coordinates (meaning not in vector notation). You will get eight equations. For example, one of them will be*

$$\frac{\partial}{\partial y} E_3 - \frac{\partial}{\partial z} E_2 = -\frac{\partial}{\partial t} B_1.$$

Exercise 2.3.5. *Let $c = 1$. Show that*

$$E = (y - z, -2zt, -x - z^2)$$
$$B = (-1 - t^2, 0, 1 + t)$$
$$\rho = -2z$$
$$j = (0, 2z, 0)$$

satisfy Maxwell's equations.

Comment: In the real world, the function ρ and the vector fields E, B, and j are determined from experiment. That is not how I chose the function and vector fields in problem 5. In Chapter 7, we will see that given any function $\phi(x,y,z,t)$ and vector field $A(x,y,z,t) = (A_1, A_2, A_3)$, if we set

$$E = -\nabla(\phi) - \frac{\partial A}{\partial t}$$
$$= \left(\frac{\partial\phi}{\partial x}, \frac{\partial\phi}{\partial y}, \frac{\partial\phi}{\partial z}\right) - \left(\frac{\partial A_1}{\partial t}, \frac{\partial A_2}{\partial t}, \frac{\partial A_3}{\partial t}\right)$$
$$B = \nabla \times A$$

and, using these particular E and B, set

$$\rho = \nabla \cdot E$$
$$j = \nabla \times B - \frac{\partial B}{\partial t},$$

we will have that ρ, E, B, and j satisfy Maxwell's equations. For this last problem, I simply chose, almost at random, $\phi(x,y,z,t) = xz$ and

$$A = (-yt + x^2, x + zt^2, -y + z^2t).$$

The punchline of Chapter 7 is that the converse holds, meaning that if the function ρ and the vector fields E, B, and j satisfy Maxwell's equations, then there must be a function $\phi(x,y,z,t)$ and a vector field A such that $E = -\nabla(\phi) - \frac{\partial A}{\partial t}$ and $B = \nabla \times A$. The $\phi(x,y,z,t)$ and A are called the *potentials*.

3
Electromagnetic Waves

Summary: When the current j and the density ρ are zero, both the electric field and the magnetic field satisfy the wave equation, meaning that both fields can be viewed as waves. In the first section, we will review the wave equation. In the second section, we will see why Maxwell's equations yield these electromagnetic waves, each having speed c.

3.1. The Wave Equation

Waves permeate the world. Luckily, there is a class of partial differential equations (PDEs) whose solutions describe many actual waves. We will not justify why these PDEs describe waves but instead will just state their form. (There are many places to see heuristically why these PDEs have anything at all to do with waves; for example, see [26].)

The one-dimensional wave equation is

$$\frac{\partial^2 y}{\partial t^2} - v^2 \frac{\partial^2 y}{\partial x^2} = 0.$$

Here the goal is to find a function $y = y(x,t)$, where x is position and t is time, that satisfies the preceding equation. Thus the "unknown" is the function $y(x,t)$. For a fixed t, this can describe a function that looks like Figure 3.1.

From the heuristics of the derivation of this equation, the speed of this wave is v.

The two-dimensional wave equation is

$$\frac{\partial^2 z}{\partial t^2} - v^2 \left(\frac{\partial^2 z}{\partial x^2} + \frac{\partial^2 z}{\partial y^2} \right) = 0.$$

Here, the function $z(x, y, t)$ is the unknown, where x and y describe position and t is again time. This could model the motion of a wave over the (x, y)-plane. This wave also has speed v.

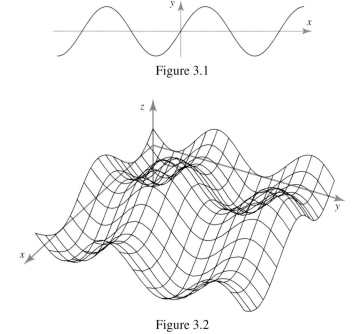

Figure 3.1

Figure 3.2

The three-dimensional wave equation is

$$\frac{\partial^2 f}{\partial t^2} - v^2 \left(\frac{\partial^2 f}{\partial x^2} + \frac{\partial^2 f}{\partial y^2} + \frac{\partial^2 f}{\partial z^2} \right) = 0.$$

Once again, the speed is v.

Any function that satisfies such a PDE is said to satisfy the wave equation. We expect such functions to have wave-like properties.

The sum of second derivatives $\frac{\partial^2 f}{\partial x^2} + \frac{\partial^2 f}{\partial y^2} + \frac{\partial^2 f}{\partial z^2}$ occurs often enough to justify its own notation and its own name, the *Laplacian*. We use the notation

$$\Delta(f) = \frac{\partial^2 f}{\partial x^2} + \frac{\partial^2 f}{\partial y^2} + \frac{\partial^2 f}{\partial z^2}.$$

This has a convenient formulation using the nabla notation $\nabla = (\frac{\partial}{\partial x}, \frac{\partial}{\partial y}, \frac{\partial}{\partial z})$. Thinking of ∇ as a vector, we interpret its dot product with itself as

$$
\begin{aligned}
\nabla^2 &= \nabla \cdot \nabla \\
&= \left(\frac{\partial}{\partial x}, \frac{\partial}{\partial y}, \frac{\partial}{\partial z} \right) \cdot \left(\frac{\partial}{\partial x}, \frac{\partial}{\partial y}, \frac{\partial}{\partial z} \right) \\
&= \frac{\partial^2}{\partial x^2} + \frac{\partial^2}{\partial y^2} + \frac{\partial^2}{\partial z^2},
\end{aligned}
$$

and thus we have

$$
\nabla^2 = \triangle.
$$

Then we can write the three-dimensional wave equation in the following three ways:

$$
\begin{aligned}
0 &= \frac{\partial^2 f}{\partial t^2} - v^2 \left(\frac{\partial^2 f}{\partial x^2} + \frac{\partial^2 f}{\partial y^2} + \frac{\partial^2 f}{\partial z^2} \right) \\
&= \frac{\partial^2 f}{\partial t^2} - v^2 \triangle(f) \\
&= \frac{\partial^2 f}{\partial t^2} - v^2 \nabla^2(f).
\end{aligned}
$$

But this wave equation is for functions $f(x, y, z, t)$. What does it mean for a vector field $F = (F_1, F_2, F_3)$ to satisfy a wave equation? We will say that the vector field F satisfies the wave equation

$$
\frac{\partial^2 F}{\partial t^2} - v^2 \triangle(F) = 0,
$$

if the following three partial differential equations

$$
\frac{\partial^2 F_1}{\partial t^2} - v^2 \left(\frac{\partial^2 F_1}{\partial x^2} + \frac{\partial^2 F_1}{\partial y^2} + \frac{\partial^2 F_1}{\partial z^2} \right) = 0
$$

$$
\frac{\partial^2 F_2}{\partial t^2} - v^2 \left(\frac{\partial^2 F_2}{\partial x^2} + \frac{\partial^2 F_2}{\partial y^2} + \frac{\partial^2 F_2}{\partial z^2} \right) = 0
$$

$$
\frac{\partial^2 F_3}{\partial t^2} - v^2 \left(\frac{\partial^2 F_3}{\partial x^2} + \frac{\partial^2 F_3}{\partial y^2} + \frac{\partial^2 F_3}{\partial z^2} \right) = 0
$$

hold.

3.2. Electromagnetic Waves

In the half-century before Maxwell wrote down his equations, an amazing amount of experimental work on the links between electricity and magnetism had been completed. To some extent, Maxwell put these empirical observations into a more precise mathematical form. These equations's strength, though, is reflected in that they allowed Maxwell, for purely theoretical reasons, to make one of the most spectacular intellectual leaps ever: Namely, Maxwell showed that electromagnetic waves that move at the speed c had to exist. Maxwell knew that this speed c was the same as the speed of light, leading him to predict that light was just a special type of elecromagnetic wave. No one before Maxwell realized this. In the 1880s, Hertz proved experimentally that light was indeed an electromagnetic wave.

We will first see intuitively why Maxwell's equations lead to the existence of electromagnetic waves, and then we will rigorously prove this fact. Throughout this section, assume that there is no charge ($\rho = 0$) and no current ($j = 0$) (i.e., we are working in a vacuum). Then Maxwell's equations become

$$\nabla \cdot E = 0$$
$$\nabla \times E = -\frac{\partial B}{\partial t}$$
$$\nabla \cdot B = 0$$
$$c^2 \, \nabla \times B = \frac{\partial E}{\partial t}.$$

These vacuum equations are themselves remarkable since they show that the waves of the electromagnetic field move in space without any charge or current.

Suppose we change the magnetic field (possibly by moving a magnet). Then $\frac{\partial B}{\partial t} \neq 0$ and hence the electric vector field E will have non-zero curl. Thus E will have a change in a direction perpendicular to the change in B. But then $\frac{\partial E}{\partial t} \neq 0$, creating curl in B, which, in turn, will prevent $\frac{\partial B}{\partial t}$ from being zero, starting the whole process over again, never stopping.

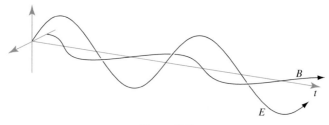

Figure 3.3

This is far from showing that we have an actual wave, though.

Now we show that the electric field E satisfies the wave equation

$$\frac{\partial^2 E}{\partial t^2} - c^2 \triangle(E) = 0.$$

We have

$$\frac{\partial^2 E}{\partial t^2} = \frac{\partial}{\partial t}\left(\frac{\partial E}{\partial t}\right)$$

$$= \frac{\partial}{\partial t}(c^2 \, \nabla \times B)$$

$$= c^2 \, \nabla \times \frac{\partial B}{\partial t}$$

$$= -c^2 \, \nabla \times \nabla \times E$$

$$= c^2 \, (\triangle(E_1), \triangle(E_2), \triangle(E_3)),$$

which means that the electric field E satisfies the wave equation. Note that the second and fourth lines result from Maxwell's equations. The fact that $\nabla \times \nabla \times E = -(\triangle(E_1), \triangle(E_2), \triangle(E_3))$ is a calculation coupled with Maxwell's first equation, which we leave for the exercises. The justification for the third equality, which we also leave for the exercises, stems from the fact that the order of taking partial derivatives is interchangeable. The corresponding proof for the magnetic field is similar and is also left as an exercise.

3.3. The Speed of Electromagnetic Waves Is Constant

3.3.1. Intuitive Meaning

We have just seen there are electromagnetic waves moving at speed c, when there is no charge and no current. In this section we want to start seeing that the existence of these waves, moving at that speed c, strikes a blow to our common-sense notions of physics, leading, in the next chapter, to the heart of the Special Theory of Relativity.

Consider a person A. She thinks she is standing still. A train passes by, going at the constant speed of 60 miles per hour. Let person B be on the train. B legitimately can think that he is at the origin of the important coordinate system, thus thinking of himself as standing still. On this train, B rolls a ball forward at, say, 3 miles per hour, with respect to the train. Observer A, though, would say that the ball is moving at $3 + 60$ miles per hour. So far, nothing controversial at all.

Let us now replace the ball with an electromagnetic wave. Suppose person B turns it on and observes it moving in the car. If you want, think of B as turning on a flashlight. B will measure its speed as some c miles per hour. Observer A will also see the light. Common sense tells us, if not screams at us, that A will measure the light as traveling at $c + 60$ miles per hour.

But what do Maxwell's equations tell us? The speed of an electromagnetic wave is the constant c that appears in Maxwell's equations. But the value of c does not depend on the initial choice of coordinate system. The (x, y, z, t) for person A and the (x, y, z, t) for person B have the same c in Maxwell. Of course, the number c in the equations is possibly only a "constant" once a coordinate system is chosen. If this were the case, then if person A measured the speed of an electromagnetic wave to be some c, then the corresponding speed for person B would be $c - 60$, with this number appearing in person B's version of Maxwell's equations. This is now an empirical question about the real world. Let A and B each measure the speed of an electromagnetic wave. What physicists find is that for both observers the speed is the same. Thus, in the preceding train, the speed of an electromagnetic wave for both A and B is the same c. This is truly bizarre.

3.3.2. Changing Coordinates for the Wave Equation

Suppose we again have two people, A and B. Let person B be traveling at a constant speed α with respect to A, with A's and B's coordinate systems exactly matching up at time $t = 0$.

To be more precise, we think of person A as standing still, with coordinates x' for position and t' for time, and of person B as moving to the right at speed α, with position coordinate x and time coordinate t. If the two coordinate systems line up at time $t = t' = 0$, then classically we would expect

$$x' = x + \alpha t$$
$$t' = t,$$

or, equivalently,

$$x = x' - \alpha t$$
$$t = t'.$$

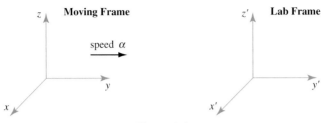

Figure 3.4

This reflects our belief that the way a stopwatch works should not be influenced by how fast it is moving. (This belief will also be shattered by the Special Theory of Relativity.)

Suppose in the reference frame for B we have a wave $y(x,t)$ satisfying

$$\frac{\partial^2 y}{\partial t^2} - v^2 \frac{\partial^2 y}{\partial x^2} = 0.$$

In B's reference frame, the speed of the wave is v. From calculus, this speed v must be equal to the rate of change of x with respect to t, or in other words $v = dx/dt$. This in turn forces

$$\frac{\partial y}{\partial t} = -v \frac{\partial y}{\partial x},$$

as we will see now. Fix a value of $y(x,t) = y_0$. Then we have some x_0 such that

$$y_0 = y(x_0, 0).$$

This means that for all x and t such that $x_0 = x - vt$, then $y_0 = y(x,t)$. The speed of the wave is how fast this point with y-coordinate y_0 moves along the x-axis with respect to t. Then

$$
\begin{aligned}
0 &= \frac{dy}{dt} \\
&= \frac{\partial y}{\partial x} \frac{dx}{dt} + \frac{\partial y}{\partial t} \frac{dt}{dt} \\
&= \frac{\partial y}{\partial x} \frac{dx}{dt} + \frac{\partial y}{\partial t} \\
&= v \frac{\partial y}{\partial x} + \frac{\partial y}{\partial t},
\end{aligned}
$$

giving us that $\partial y/\partial t = -v \partial y/\partial x$.

Person A is looking at the same wave but measures the wave as having speed $v + \alpha$. We want to see explicitly that under the appropriate change of coordinates this indeed happens. This is an exercise in the chain rule, which is critically important in these arguments.

Our wave $y(x,t)$ can be written as function of x' and t', namely, as

$$y(x,t) = y(x' - \alpha t', t').$$

We want to show that this function satisfies

$$\frac{\partial^2 y}{\partial t'^2} - (v + \alpha)^2 \frac{\partial^2 y}{\partial x'^2} = 0.$$

The key will be that

$$\frac{\partial}{\partial x'} = \frac{\partial x}{\partial x'}\frac{\partial}{\partial x} + \frac{\partial t}{\partial x'}\frac{\partial}{\partial t}$$

$$= \frac{\partial}{\partial x}$$

$$\frac{\partial}{\partial t'} = \frac{\partial x}{\partial t'}\frac{\partial}{\partial x} + \frac{\partial t}{\partial t'}\frac{\partial}{\partial t}$$

$$= -\alpha\frac{\partial}{\partial x} + \frac{\partial}{\partial t}$$

by the chain rule.

We start by showing that

$$\frac{\partial y}{\partial t'} = -\alpha\frac{\partial y}{\partial x} + \frac{\partial y}{\partial t}$$

and

$$\frac{\partial y}{\partial x'} = \frac{\partial y}{\partial x},$$

whose proofs are left for the exercises.

Turning to second derivatives, we can similarly show that

$$\frac{\partial^2 y}{\partial t'^2} = \alpha^2\frac{\partial^2 y}{\partial x^2} - 2\alpha\frac{\partial^2 y}{\partial x\partial t} + \frac{\partial^2 y}{\partial t^2}$$

and

$$\frac{\partial^2 y}{\partial x'^2} = \frac{\partial^2 y}{\partial x^2},$$

whose proofs are also left for the exercises.

Knowing that $\frac{\partial y}{\partial t} = -v\frac{\partial y}{\partial x}$, we have

$$\frac{\partial^2 y}{\partial x\partial t} = -v\frac{\partial^2}{\partial x^2}.$$

This allows us to show (left again for the exercises) that

$$\frac{\partial^2 y}{\partial t'^2} - (v + \alpha)^2 \frac{\partial^2 y}{\partial x'^2} = 0,$$

which is precisely what we desired.

Classically, the speed of a wave depends on the coordinate system that is being used. Maxwell's equations tell us that this is not true, at least for electromagnetic waves. Either Maxwell or the classical theory must be wrong. In the next chapter, we will explore the new theory of change of coordinates implied by Maxwell's equations, namely, the Special Theory of Relativity.

3.4. Exercises

Exercise 3.4.1. *Show that the functions $y_1(x,t) = \sin(x - vt)$ and $y_2(x,t) = \sin(x + vt)$ are solutions to the wave equation*

$$\frac{\partial^2 y}{\partial t^2} - v^2 \frac{\partial^2 y}{\partial x^2} = 0.$$

Exercise 3.4.2. *Let $f(u)$ be a twice differentiable function. Show that both $f(x - vt)$ and $f(x + vt)$ are solutions to the wave equation. Then interpret the solution $f(x - vt)$ as the graph $y = f(u)$ moving to the right at a speed v and the solution $f(x + vt)$ as the graph $y = f(u)$ moving to the left at a speed v.*

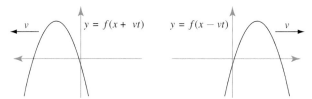

Figure 3.5

Exercise 3.4.3. *Suppose that $f_1(x,t)$ and $f_2(x,t)$ are solutions to the wave equation $\frac{\partial^2 f}{\partial t^2} - v^2 \triangle(f) = 0$. Show that $\lambda_1 f_1(x,t) + \lambda_2 f_2(x,t)$ is another solution, where λ_1 and λ_2 are any two real numbers.*

Comment: When a differential equation has the property that $\lambda_1 f_1(x,t) + \lambda_2 f_2(x,t)$ is a solution whenever $f_1(x,t)$ and $f_2(x,t)$ are solutions, we say that the differential equation is *homogeneous linear*.

Exercise 3.4.4. *Show that*

$$\frac{\partial}{\partial t}(\nabla \times F) = \nabla \times \frac{\partial F}{\partial t}.$$

Exercise 3.4.5. *Let $B(x,y,z,t)$ be a magnetic field when there is no charge ($\rho = 0$) and no current ($j = 0$). Show that B satisfies the wave equation $\frac{\partial^2 B}{\partial t^2} - c^2 \triangle(B) = 0$. First prove this following the argument in the text. Then close this book and recreate the argument from memory. Finally, in a few hours, go through the argument again.*

Exercise 3.4.6. *Suppose that* $\nabla \cdot E = 0$. *Show that*

$$\nabla \times \nabla \times E = -(\triangle(E_1), \triangle(E_2), \triangle(E_3)).$$

As a word of warning, this is mostly a long calculation, though at a critical point you will need to use that $\nabla \cdot E = 0$.

Exercise 3.4.7. *Let*

$$E = (0, \sin(x - ct), 0).$$

Assume that there is no charge ($\rho = 0$) and no current ($j = 0$). Find a corresponding magnetic field B and then show that both E and B satisfy the wave equation.

Exercise 3.4.8. *Using the notation from Section 3.3, show that*

$$\frac{\partial y}{\partial t'} = -\alpha \frac{\partial y}{\partial x} + \frac{\partial y}{\partial t}$$

and

$$\frac{\partial y}{\partial x'} = \frac{\partial y}{\partial x}.$$

Exercise 3.4.9. *Using the notation from Section 3.3, show that*

$$\frac{\partial^2 y}{\partial t'^2} = \alpha^2 \frac{\partial^2 y}{\partial x^2} - 2\alpha \frac{\partial^2 y}{\partial x \partial t} + \frac{\partial^2 y}{\partial t^2}$$

and

$$\frac{\partial^2 y}{\partial x'^2} = \frac{\partial^2 y}{\partial x^2}.$$

Exercise 3.4.10. *Using the notation from Section 3.3, show that*

$$\frac{\partial^2 y}{\partial t'^2} - (v + \alpha)^2 \frac{\partial^2 y}{\partial x'^2} = 0.$$

(Since we know that $\frac{\partial y}{\partial t} = -v \frac{\partial y}{\partial x}$, *we know that*

$$\frac{\partial^2 y}{\partial x \partial t} = -v \frac{\partial^2 y}{\partial x^2}.)$$

4

Special Relativity

Summary: We develop the basics of special relativity. Key is determining the allowable coordinate changes. This will let us show in the next chapter not just how but why magnetism and electricity must be linked.

4.1. Special Theory of Relativity

The physics and mathematics of Maxwell's equations in the last chapter were worked out during the 1800s. In these equations, there is no real description for why electricity and magnetism should be related. Instead, the more practical description of the how is treated. Then, in 1905, came Albert Einstein's "On the Electrodynamics of Moving Bodies" [19], the paper that introduced the world to the Special Theory of Relativity. This paper showed the why of Maxwell's equations (while doing far more).

The Special Theory of Relativity rests on two assumptions, neither at first glance having much to do with electromagnestism.

Assumption I: *Physics must be the same in all frames of reference moving at constant velocities with respect to each other.*

Assumption II: *The speed of light in a vacuum is the same in all frames of reference that move at constant velocities with respect to each other.*

The first assumption is quite believable, saying in essence that how you choose your coordinates should not affect the underlying physics of what you are observing. It leads to an interesting problem, though, of how to translate from one coordinate system to another. It is the second assumption that will drastically restrict how we are allowed to change coordinate systems.

It is also this second assumption that is, at first glance, completely crazy. In the last chapter we saw that this craziness is already hidden in Maxwell's

equations. Assumption II is an empirical statement, one that can be tested. In
all experiments ever done (and there have been many), Assumption II holds.
The speed of light is a constant.

But Einstein in 1905 had not done any of these experiments. How did he
ever come up with such an idea? The answer lies in Maxwell's equations.
As we have seen, Maxwell's equations give rise to electromagnetic waves
that move at c, where c is the speed of light. But this speed is independent
of the observer. The electromagnetic waves in Maxwell's equations satisfy
Assumption II. This fact wreaks havoc on our common sense notions of space
and time.

4.2. Clocks and Rulers

To measure anything, we must first choose units of measure. For now, we
want to describe where something happens (its spatial coordinates) and when
it happens (its time). Fixing some point in space as the origin and some time
as time zero, choosing units of length and of time, and finally choosing three
axes in space, we see that we can describe any event in terms of four numbers
(x, y, z, t), with (x, y, z) describing the position in space and t the time.

Suppose we set up two coordinate systems. The first we will call the
laboratory frame, denoting its coordinates by

$$(x_1, y_1, z_1, t_1).$$

Think of this frame as motionless. Now suppose we have another frame of
reference, which we will call the moving frame. Its coordinates are denoted by

$$(x_2, y_2, z_2, t_2).$$

Suppose this new frame is moving along at a constant speed of v units/sec in
a direction parallel to the x-axis, with respect to the lab frame. Further we
suppose that the two frames agree at some specified event.

You can think of the moving frame as being in a train that is chugging along
in the lab.

Let c be the speed of light. Assumption II states that this speed c must be
the same in both reference frames. Suppose we put a mirror in the moving
frame at the point

$$x_2 = 0$$

$$y_2 = c$$

$$z_2 = 0$$

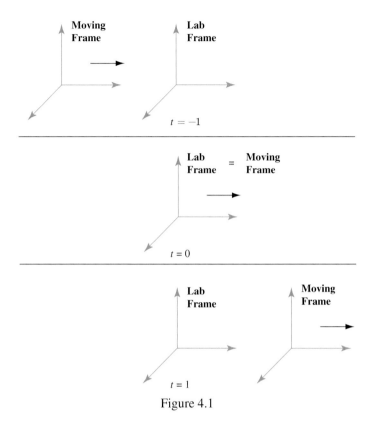

Figure 4.1

At $t = 0$, shine a flashlight, in the moving frame, from the moving frame's origin toward the mirror. (The flashlight is stationary in the moving frame but appears to be moving at speed v in the stationary frame.) The light bounces off the mirror, right back to the moving frame's origin.

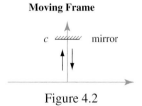

Figure 4.2

It takes one second for the light to go from the origin to the mirror and another second to return. Thus the light will travel a distance of $2c$ units over a time of 2 seconds.

Thus the velocity of light will be what we expect:

$$\text{speed of light} = \frac{\text{distance}}{\text{time}} = \frac{2c}{2} = c.$$

What does someone in the lab frame see? In the lab frame, at time $t = 0$, suppose that the flashlight is turned on at the place

$$x_1 = -v$$
$$y_1 = 0$$
$$z_1 = 0$$

At this time, the mirror is at

$$x_1 = -v$$
$$y_1 = c$$
$$z_1 = 0$$

The light will appear to move along this path

Lab Frame

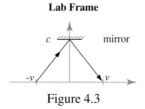

Figure 4.3

in the lab frame. Common sense tells us that the light should still take 2 seconds to complete its path, a path of length $2\sqrt{v^2 + c^2}$. It appears that the speed of light in the lab frame should be

$$\text{speed of light} = \frac{\text{distance}}{\text{time}} = \frac{2\sqrt{v^2 + c^2}}{2} = \sqrt{v^2 + c^2}.$$

This cannot happen under Assumption II.

This leads to the heart of the problem. We need to know how to translate from one of our coordinate systems to the other. Assumption II will require us to allow surprising and non-intuitive translations.

4.3. Galilean Transformations

There is no mystery that people use different coordinate systems in different places and times. Special relativity highlights, though, that how one coordinate system can be changed into another is at the heart of mechanics.

Here we will write the classical change of coordinates (called Galilean transformations) via matrix transformations. Later in the chapter, we will describe the analogous transformations for special relativity (the Lorentz transformations), also via matrices.

We start with two reference frames, one moving with speed v with respect to the other. We will just follow our nose and ignore any condition on the speed of light.

Assume, as would seem obvious, that we can synchronize the times in the two frames, meaning that

$$t_1 = t_2.$$

Suppose that at time zero, the two origins are at the same point but that one frame is moving in the x-direction with speed v with respect to the other frame. Then the spatial coordinates are related by:

$$x_2 = x_1 + vt$$

$$y_2 = y_1$$

$$z_2 = z_1.$$

We are in a Newtonian world, where the origin in the moving frame is moving along the x_1-axis in the lab frame. (As we saw earlier, this type of change of coordinates violates the assumption that the speed of light must be constant in all frames of reference.) We can write this change of coordinates as the matrix transformation

$$\begin{pmatrix} x_2 \\ y_2 \\ z_2 \\ t_2 \end{pmatrix} = \begin{pmatrix} 1 & 0 & 0 & v \\ 0 & 1 & 0 & 0 \\ 0 & 0 & 1 & 0 \\ 0 & 0 & 0 & 1 \end{pmatrix} \begin{pmatrix} x_1 \\ y_1 \\ z_1 \\ t_1 \end{pmatrix}.$$

This format suggests how to write down changes of coordinates in general. We keep with the assumption that the times in both coordinate systems are equal and synchronized ($t_1 = t_2$) and that at time zero, the two origins are at the same point. We further assume that a unit of length in one reference frame is the same as a unit of length in the other and that angles are measured the same. Then all allowable changes of coordinates take the form

$$
\begin{pmatrix} x_2 \\ y_2 \\ z_2 \\ t_2 \end{pmatrix} = \begin{pmatrix} a_{11} & a_{12} & a_{13} & v_1 \\ a_{21} & a_{22} & a_{23} & v_2 \\ a_{31} & a_{32} & a_{33} & v_3 \\ 0 & 0 & 0 & 1 \end{pmatrix} \begin{pmatrix} x_1 \\ y_1 \\ z_1 \\ t_1 \end{pmatrix}.
$$

In order for the lengths and angles to be preserved in the two reference frames, we need the dot products between vectors in the two frames to be preserved when transformed. This means that we need the matrix

$$
A = \begin{pmatrix} a_{11} & a_{12} & a_{13} \\ a_{21} & a_{22} & a_{23} \\ a_{31} & a_{32} & a_{33} \end{pmatrix}
$$

to be in the orthogonal group $O(3, \mathbb{R})$, which is just a fancy way of saying the following:

Recall that the transpose of the matrix A is

$$
A^T = \begin{pmatrix} a_{11} & a_{21} & a_{31} \\ a_{12} & a_{22} & a_{32} \\ a_{13} & a_{23} & a_{33} \end{pmatrix}.
$$

Then $A \in O(3, \mathbb{R})$ means that for all vectors $(x \quad y \quad z)$ and $(u \quad v \quad w)$, we have

$$
(x \quad y \quad z) \cdot \begin{pmatrix} u \\ v \\ w \end{pmatrix} = (x \quad y \quad z) A^T A \begin{pmatrix} u \\ v \\ w \end{pmatrix}.
$$

These orthogonal transformations are the only allowable change of coordinates for transforming from one frame of reference to another frame that is moving with velocity (v_1, v_2, v_3) with respect to the first, under the assumption that the time coordinate is "independent" of the spatial coordinates.

4.4. Lorentz Transformations

4.4.1. A Heuristic Approach

(While the derivation that follows is standard, I have followed the presentation in chapter 3 of A. P. French's *Special Relativity* [25], a source I highly recommend.)

The Lorentz transformations are the correct type of changes of coordinates that will satisfy the Special Relativity requirement that the speed of light is a constant. H. A. Lorentz wrote down these coordinate changes in his papers

"Michelson's Interference Experiment" and "Electromagnetic Phenomena in a System Moving with any Velocity Less than that of Light," both reprinted in [19] and both written before Einstein's Special Theory of Relativity, in an attempt to understand the Michelson-Morley experiments (which can now be interpreted as experiments showing that the speed of light is independent of reference frame).

Suppose, as before, that we have two frames: one we call the lab frame (again with coordinates labeled (x_1, y_1, z_1, t_1)) and the other the moving frame (with coordinates (x_2, y_2, z_2, t_2)). Suppose this second frame moves at a constant speed of v units/sec in a direction parallel to the x-axis, with respect to the lab frame. Further, we suppose that the two frames coincide at time $t_1 = t_2 = 0$. We now assume that the speed of light is constant in every frame of reference and hence is a well-defined number, to be denoted by c. We want to justify that the two coordinates of the two systems are related by

$$x_2 = \gamma x_1 - \gamma v t_1$$

$$y_2 = y_1$$

$$z_2 = z_1$$

$$t_2 = -\gamma \left(\frac{v}{c^2}\right) x_1 + \gamma t_1,$$

where

$$\gamma = \frac{1}{\sqrt{1 - (\frac{v}{c})^2}}.$$

It is certainly not clear why this is at all reasonable. But if this coordinate change is correct, then a major problem will occur if the velocity is greater than the speed of light, as γ will then be an imaginary number. Hence, if the previous transformations are indeed true, we can conclude that the speed of light must be an upper bound for velocity.

We can rewrite the preceding change of coordinates in the following matrix form:

$$\begin{pmatrix} x_2 \\ y_2 \\ z_2 \\ t_2 \end{pmatrix} = \begin{pmatrix} \gamma & 0 & 0 & -v\gamma \\ 0 & 1 & 0 & 0 \\ 0 & 0 & 1 & 0 \\ \gamma(-\frac{v}{c^2}) & 0 & 0 & \gamma \end{pmatrix} \begin{pmatrix} x_1 \\ y_1 \\ z_1 \\ t_1 \end{pmatrix}.$$

We now give a heuristic justification for these transformations, critically using both of our assumptions for special relativity.

First, since the two reference frames are only moving with respect to each other along their x-axes, it is reasonable to assume that the two coordinate systems agree in the y and z directions, motivating our letting $y_2 = y_1$ and

$z_2 = z_1$, which in turn allows us to ignore the y and z coordinates. We make the additional assumption that there are constants a and b such that

$$x_2 = ax_1 + bt_1.$$

Thus, we are explicitly assuming that any change of coordinates should be linear. Now consider Assumption I. An observer in frame 1 thinks that frame 2 is moving to the right at speed v. But an observer in frame 2 will think of themselves as standing still, with frame 1 whizzing by to the left at speed v. Assumption I states that there is intrinsically no way for anyone to be able to distinguish the two frames, save for the directions they are moving with respect to each other. In particular, neither observer can claim to be the frame that is "really" standing still. This suggests that

$$x_1 = ax_2 - bt_2$$

with the minus sign in front of the b coefficient reflecting that the two frames are moving in opposite directions. Here we are using Assumption I, namely, that we cannot distinguish the two frames, save for how they are moving with respect to each other.

Now to start getting a handle on a and b. Consider how frame 2's origin $x_2 = 0$ is moving in frame 1. To someone in frame 1, this point must be moving at speed v to the right. Since $x_2 = ax_1 + bt_1$, if $x_2 = 0$, we must have $x_1 = -\left(\frac{b}{a}\right)t_1$ and hence

$$v = \frac{dx_1}{dt_1} = -\frac{b}{a}.$$

Thus we have

$$x_2 = ax_1 - avt_1$$

$$x_1 = ax_2 + avt_2.$$

Assumption II now moves to the fore. The speed of light is the same in all reference frames. Suppose we turn on a flashlight at time $t_1 = t_2 = 0$ at the origin of each frame, shining the light along each frame's x-axis. Then we have, in both frames, that the path of the light is described by

$$x_1 = ct_1$$

$$x_2 = ct_2.$$

We have

$$ct_2 = x_2 = ax_1 - avt_1 = act_1 - avt_1 = a(c - v)t_1$$

and

$$ct_1 = x_1 = ax_2 + avt_2 = act_2 + avt_2 = a(c + v)t_2.$$

Dividing through by c in the second equation and then plugging in for t_1 in the first equation, we get

$$ct_2 = \left(\frac{a^2(c^2 - v^2)}{c} \right) t_2$$

which means that

$$a^2 = \frac{1}{1 - (\frac{v}{c})^2}$$

finally yielding that

$$a = \frac{1}{\sqrt{1 - (\frac{v}{c})^2}} = \gamma.$$

We leave the proof that

$$t_2 = -\gamma \left(\frac{v}{c^2} \right) x_1 + \gamma t_1$$

as an exercise at the end of the chapter.

4.4.2. Lorentz Contractions and Time Dilations

(This section closely follows [25].)

The goal of this section is to show how length and time changes are not invariant but depend on the frame in which they are measured. Suppose you are in a reference frame. Naturally, you think you are standing still. A friend whizzes by, in the positive x direction, at velocity v. Each of you has a meter stick and a stopwatch. As she moves by you, you notice that her meter stick is shorter than yours and her second hand is moving more slowly than your second hand. You call for her to stop and stand next to you and discover that now the two meter sticks exactly line up and that the two stopwatches tick exactly the same. Something strange is going on. In this section, we use the previous Lorentz transformations to explain this weirdness in measurements of length and time.

We start with length. Suppose we measure, in our reference frame, the distance between a point at x_1 and a point at x_2 with both points on the x-axis. The length is simply

$$l = x_2 - x_1,$$

assuming that $x_1 < x_2$. But what is this in the moving frame? We have to make the measurement in the moving frame when the time is constant. So suppose in the moving frame that the point at x_1 is at x_1' at time t' and the point at x_2 is at x_2' at time t'. In the moving frame the distance between the points is

$$l' = x_2' - x_1'.$$

Via the Lorentz transformations, we can translate between the coordinates in the two frames, using that $x = \gamma x' + v\gamma t'$ where $\gamma = \dfrac{1}{\sqrt{1-(v/c)^2}}$. Then we have

$$x_2 = \gamma x_2' + v\gamma t'$$

$$x_1 = \gamma x_1' + v\gamma t'.$$

Hence

$$l = x_2 - x_1 = \gamma(x_2' - x_1') = \gamma l'.$$

Lengths must vary.

Now to see that time must also vary. Here we fix a position in space and make two time measurements. In our frame, where we think we are standing still, suppose at some point x we start a stopwatch at time t_1 and stop it at t_2. We view the elapsed time as simply

$$\Delta t = t_2 - t_1.$$

Let the times in the moving frame be denoted by t_1' and t_2', in which case the change in time in this frame is

$$\Delta t' = t_2' - t_1'.$$

Using that $t' = \gamma\left(\dfrac{v}{c^2}\right)x + \gamma t$, we have that

$$t_1' = \gamma\left(\frac{v}{c^2}\right)x + \gamma t_1$$

$$t_2' = \gamma\left(\frac{v}{c^2}\right)x + \gamma t_2,$$

which gives us

$$\Delta t' = t_2' - t_1' = \gamma(t_2 - t_1) = \gamma \Delta t.$$

Note that since $v < c$, we have $\gamma > 1$. Thus $l' < l$ while $\Delta t' > \Delta t$, which is why the we use the terms "Lorentz contraction" and "time dilation."

4.4.3. Proper Time

There is no natural coordinate system. There is no notion of "true" time or "true" position. Still, for a given particle moving at a constant velocity, there is one measure of time that can be distinguished from all others, namely, the time on a coordinate system when the particle's velocity is simply zero. Thus, for this particular coordinate system, the particle stays at the origin for all time. The measure of time for this coordinate system has its own name: *proper time*.

From the previous subsection, if we use $t, x, y,$ and z to denote the coordinate system for which a particle remains at the origin, and $t', x', y',$ and

z' to denote a coordinate system moving with speed v with respect to the first coordinate system, we have

$$\Delta t' = t_2' - t_1' = \frac{\Delta t}{\sqrt{1 - (\frac{v}{c})^2}} = \gamma(t_2 - t_1) = \gamma \Delta t.$$

The use of proper time will be critical when we define the relativistic version of momentum.

4.4.4. The Special Relativity Invariant

Here is how the world works. Start with two friends, both having identical meter sticks and identical stopwatches. Now suppose each is moving at a speed v with respect to the other. (Thus both think that they are perfectly still and it is their friend who is moving.) Both think that their friend's meter stick has shrunk while their friend's stopwatch is running slow. The very units of length and time depend on which reference frame you are in. These units are not invariants but are *relative* to what reference frame they are measured in. This is why this theory is called the Special Theory of Relativity.

This does not mean that everything is relative. Using Lorentz transformations, we can translate the lengths and times in one reference frame to those in another. But there is a quantity that cannot change, no matter the reference frame. There is a number that is an *invariant*.

Suppose we have a coordinate system (x_1, y_1, z_1, t_1), where (x_1, y_1, z_1) are the three space coordinates and t is the time coordinate. Let there be another coordinate system moving to the right of the first, in the x-direction, at speed v, with corresponding coordinates (x_2, y_2, z_2, t_2). We know that

$$x_2 = \gamma x_1 - \gamma v t_1$$

$$y_2 = y_1$$

$$z_2 = z_1$$

$$t_2 = -\gamma \left(\frac{v}{c^2}\right) x_1 + \gamma t_1,$$

where

$$\gamma = \frac{1}{\sqrt{1 - (\frac{v}{c})^2}}.$$

The number that is invariant is

$$c^2 t^2 - x^2 - y^2 - z^2.$$

One of the exercises asks you to show explicitly for the previous transformation that

$$c^2 t_2^2 - x_2^2 - y_2^2 - z_2^2 = c^2 t_1^2 - x_1^2 - y_1^2 - z_1^2.$$

Note that time is playing a different role than the space coordinates.

4.4.5. Lorentz Transformations, the Minkowski Metric, and Relativistic Displacement

We want to systematize the allowable changes of coordinates. In the previous discussion, we have a very special type of change of coordinates. While it is certainly reasonable to assume that we can always choose our space coordinates so that the relative movement between reference frames is in the x-direction, this does involve a choice.

We expect that two coordinate systems moving at a constant velocity with respect to each other are related by a four-by-four matrix A such that

$$\begin{pmatrix} t_2 \\ x_2 \\ y_2 \\ z_2 \end{pmatrix} = A \begin{pmatrix} t_1 \\ x_1 \\ y_1 \\ z_1 \end{pmatrix}$$

What we need to do is find conditions on A that are compatible with relativity theory.

As seen in the previous subsection, the key is that we want

$$c^2 (\text{change in time})^2 - (\text{change in distance})^2$$

to be a constant.

Definition 4.4.1. *The map*

$$\rho : \mathbb{R}^4 \times \mathbb{R}^4 \to \mathbb{R}$$

defined by setting

$$\rho \left(\begin{pmatrix} t_2 \\ x_2 \\ y_2 \\ z_2 \end{pmatrix}, \begin{pmatrix} t_1 \\ x_1 \\ y_1 \\ z_1 \end{pmatrix} \right) = \begin{pmatrix} t_2 & x_2 & y_2 & z_2 \end{pmatrix} \begin{pmatrix} c^2 & 0 & 0 & 0 \\ 0 & -1 & 0 & 0 \\ 0 & 0 & -1 & 0 \\ 0 & 0 & 0 & -1 \end{pmatrix} \begin{pmatrix} t_1 \\ x_1 \\ y_1 \\ z_1 \end{pmatrix}$$

$$= c^2 t_1 t_2 - x_1 x_2 - y_1 y_2 - z_1 z_2$$

is the **Minkowski** *metric. Sometimes this is also called the* **Lorentz** *metric.*

This suggests that in special relativity the allowable changes of coordinates will be given by 4×4 matrices A that satisfy

$$
A^T \begin{pmatrix} c^2 & 0 & 0 & 0 \\ 0 & -1 & 0 & 0 \\ 0 & 0 & -1 & 0 \\ 0 & 0 & 0 & -1 \end{pmatrix} A = \begin{pmatrix} c^2 & 0 & 0 & 0 \\ 0 & -1 & 0 & 0 \\ 0 & 0 & -1 & 0 \\ 0 & 0 & 0 & -1 \end{pmatrix},
$$

that is, the matrices that preserve the Minkowski metric.

More succinctly, we say:

Definition 4.4.2. *An invertible 4×4 matrix A is a* Lorentz transformation *if for all vectors $v_1, v_2 \in \mathbb{R}^4$, we have*

$$
\rho(v_2, v_1) = \rho(Av_2, Av_1).
$$

The function $\rho : \mathbb{R}^4 \times \mathbb{R}^4 \to \mathbb{R}$ is the special relativistic way of measuring "distance" and thus can be viewed as an analog of the Pythagorean theorem, which gives the distance in Euclidean space. Classically, the vector that measures a particle's movement over time, or its displacement, is given by the changes in the spatial coordinates $(\triangle x, \triangle y, \triangle z)$. Relativistically, this does not seem to be a natural vector at all. Instead, motivated by the Minkowski metric, we make the following definition:

Definition 4.4.3. *The* relativistic displacement vector *is*

$$
(c\triangle t, \triangle x, \triangle y, \triangle z).
$$

Note that the Minkowski length of the displacement vector

$$
\rho((c\triangle t, \triangle x, \triangle y, \triangle z), (c\triangle t, \triangle x, \triangle y, \triangle z))
$$

is independent of which coordinate system is chosen.

There is one significant difference between the traditional "Pythagorean" metric and our new Minkowski metric, in that we can have "negative" lengths. Consider

$$
\rho(v, v) = c^2 t^2 - x^2 - y^2 - z^2.
$$

This number can be positive, negative, or even zero.

Definition 4.4.4. *A vector $v = (t, x, y, z)$ is* space-like *if*

$$
\rho(v, v) < 0,
$$

time-like *if*

$$
\rho(v, v) > 0,
$$

and light-like *if*

$$\rho(v,v) = 0.$$

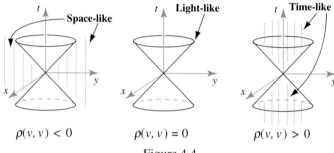

$$\rho(v,v) < 0 \qquad\qquad \rho(v,v) = 0 \qquad\qquad \rho(v,v) > 0$$

Figure 4.4

(From Figure 4.4, we can see why people say that light travels on the "light cone.") A natural question is why different vectors have the names "space-like," "time-like," and "light-like." We will now build the intuitions behind these names.

Suppose in a coordinate frame we are given two points: an event A with coordinates $(0,0,0,0)$ and an event B with coordinates $(t,0,0,0)$ with $t > 0$. We can reasonably interpret this as both A and B being at the origin of space, but with B occurring at a later time. Note that event B is time-like.

Now suppose that event B is still time-like, but now with coordinates $(t,x,0,0)$, with, say, both t and x positive. Thus we know that $c^2t^2 - x^2 > 0$ and hence that $ct > x$. We want to show that there is a Lorentz transformation leaving event A at the origin but changing event B's coordinates to $(t',0,0,0)$, meaning that in some frame of reference, moving at some velocity v with respect to the original frame, event B will seem to be standing still. Thus we must find a v such that

$$x' = \gamma x - \gamma v t = 0,$$

where $\gamma = \frac{1}{\sqrt{1-(v/c)^2}}$.

Consider the Figure 4.5:

Since B is time-like, we have

$$x = \lambda t$$

for some constant $\lambda < c$. If we set $v = \lambda$, we get our result. Thus a point is time-like if we can find a coordinate system such that the event is not moving in space.

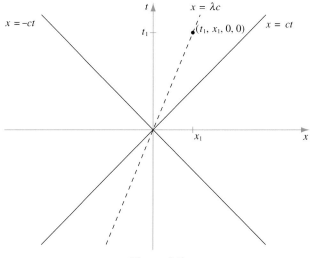

Figure 4.5

Space-like events have an analogous property. Start with our event *A* at the origin $(0,0,0,0)$, but now let event *B* have coordinates $(0,x,y,z)$. In this coordinate frame, we can interpret this as event *A* and event *B* as happening at the same time but at different places in space. Note that event *B* is space-like.

Now let event *B* still be space-like, but now with coordinates $(t,x,0,0)$, with both *t* and *x* positive. Thus we know that $ct < x$. Similarly to before, we want to find a Lorentz transformation leaving the event *A* at the origin but changing the event *B*'s coordinates to $(0,x,0,0)$, meaning that in some frame of reference, moving at some velocity *v* with respect to the original frame, the event *B* will occur at exactly the same time as event *A*. Thus we want to find a velocity *v* such that

$$t' = -\gamma \left(\frac{v}{c^2} \right) x + \gamma t = 0.$$

Since *B* is space-like, we have

$$x = \lambda t$$

for some constant $\lambda > c$.

To get $t' = 0$, we need

$$\left(\frac{v}{c^2} \right) x = t$$

or, plugging in $x = \lambda t$,

$$\left(\frac{v}{c^2} \right) \lambda t = t.$$

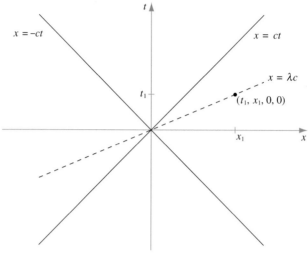

Figure 4.6

Thus we need to have

$$v = \frac{c^2}{\lambda}.$$

Here it is critical that event B is space-like, since that forces $\lambda > c$, which in turn means that v is indeed less than the speed of light c.

Consider again

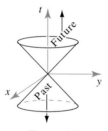

Figure 4.7

Suppose we start at the origin $(0,0,0,0)$. Is it possible for us to move to some other event with coordinates (t,x,y,z)? Certainly we need $t > 0$, but we also need (t,x,y,z) to be time-like, as otherwise we would have to move faster than the speed of light. Hence we call this region the *future*. Similarly, for an event (t,x,y,z) to be able to move to the origin, we need $t < 0$ for it to be time-like; these points are in the region called the *past*. Space-like events cannot influence the origin.

4.5. Velocity and Lorentz Transformations

Fix a coordinate frame, which we label as before with coordinates (x_1, y_1, z_1, t_1). Suppose a particle moves in this frame, with space coordinates given by functions

$$(x_1(t_1), y_1(t_1), z_1(t_1)).$$

Then the velocity vector is

$$v_1(t_1) = \left(\frac{dx_1}{dt_1}, \frac{dy_1}{dt_1}, \frac{dz_1}{dt_1} \right).$$

Now consider a different frame, with coordinates (x_2, y_2, z_2, t_2), moving with constant velocity with respect to the first frame. We know that there is a Lorentz transformation A taking the coordinates (x_1, y_1, z_1, t_1) to the coordinates (x_2, y_2, z_2, t_2). Then the moving particle must be describable via a function

$$(x_2(t_2), y_2(t_2), z_2(t_2)),$$

with velocity vector

$$v_2(t_2) = \left(\frac{dx_2}{dt_2}, \frac{dy_2}{dt_2}, \frac{dz_2}{dt_2} \right).$$

The question for this section is how to relate the vector v_1 with the vector v_2.

Since we have

$$\begin{pmatrix} t_2 \\ x_2(t_2) \\ y_2(t_2) \\ z_2(t_2) \end{pmatrix} = A \begin{pmatrix} t_1 \\ x_1(t_1) \\ y_1(t_1) \\ z_1(t_1) \end{pmatrix},$$

we can compute, for example, that

$$\frac{dx_2}{dt_2} = \frac{d(x_2 \text{ as a function of } (x_1, y_1, z_1, t_1))}{dt_1} \frac{dt_1}{dt_2},$$

allowing us to relate dx_2/dt_2 to $dx_1/dt_1, dy_1/dt_1$ and dz_1/dt_1. In a similar way, we can find dy_2/dt_2 and dz_2/dt_2 in terms of the $dx_1/dt_1, dy_1/dt_1$ and dz_1/dt_1.

To make this more concrete, let us consider one possible A, namely, the coordinate change

$$x_2 = \gamma x_1 - \gamma v t_1$$

$$y_2 = y_1$$

$$z_2 = z_1$$

$$t_2 = -\gamma \left(\frac{v}{c^2} \right) x_1 + \gamma t_1,$$

where $\gamma = 1/\sqrt{1-(\frac{v}{c})^2}$.

Since

$$\frac{dt_2}{dt_1} = \frac{d(-\gamma\left(\frac{v}{c^2}\right)x_1 + \gamma t_1)}{dt_1}$$

$$= -\gamma\left(\frac{v}{c^2}\right)\frac{dx_1}{dt_1} + \gamma$$

and

$$\frac{dx_2}{dt_1} = \frac{d(\gamma x_1 - \gamma v t_1)}{dt_1}$$

$$= \gamma\frac{dx_1}{dt_1} - \gamma v,$$

we have

$$\frac{dx_2}{dt_2} = \frac{dx_2}{dt_1}\frac{dt_1}{dt_2}$$

$$= \frac{\left(\gamma\frac{dx_1}{dt_1} - \gamma v\right)}{\left(-\gamma\left(\frac{v}{c^2}\right)\frac{dx_1}{dt_1} + \gamma\right)}$$

$$= \frac{\left(\frac{dx_1}{dt_1} - v\right)}{\left(-\left(\frac{v}{c^2}\right)\frac{dx_1}{dt_1} + 1\right)}.$$

In a similar way, as will be shown in the exercises, we have

$$\frac{dy_2}{dt_2} = \frac{\frac{dy_1}{dt_1}}{\left(-\gamma\left(\frac{v}{c^2}\right)\frac{dx_1}{dt_1} + \gamma\right)}$$

$$\frac{dz_2}{dt_2} = \frac{\frac{dz_1}{dt_1}}{\left(-\gamma\left(\frac{v}{c^2}\right)\frac{dx_1}{dt_1} + \gamma\right)}.$$

4.6. Acceleration and Lorentz Transformations

Using the notation from the last section, the acceleration vector in frame 1 is

$$a_1(t_1) = \left(\frac{d^2x_1}{dt_1^2}, \frac{d^2y_1}{dt_1^2}, \frac{d^2z_1}{dt_1^2} \right),$$

while the acceleration vector in frame 2 is

$$a_2(t_2) = \left(\frac{d^2x_2}{dt_2^2}, \frac{d^2y_2}{dt_2^2}, \frac{d^2z_2}{dt_2^2} \right).$$

We can relate the vector $a_2(t_2)$ to $a_1(t_1)$ in a similar fashion to how we related the velocities in the previous section. For example, we have

$$\frac{d^2x_2}{dt_2^2} = \frac{d}{dt_1} \left(\frac{dx_2}{dt_2} \right) \frac{dt_1}{dt_2},$$

and we know how to write dx_2/dt_2 in terms of the coordinates of frame 1. To be more specific, we again consider the Lorentz transformation

$$x_2 = \gamma x_1 - \gamma v t_1$$

$$y_2 = y_1$$

$$z_2 = z_1$$

$$t_2 = -\gamma \left(\frac{v}{c^2} \right) x_1 + \gamma t_1,$$

where $\gamma = 1/\sqrt{1 - (\frac{v}{c})^2}$. Then we have

$$\frac{d^2x_2}{dt_2^2} = \frac{d}{dt_1} \left(\frac{dx_2}{dt_2} \right) \frac{dt_1}{dt_2}$$

$$= \frac{d}{dt_1} \left(\frac{\left(\frac{dx_1}{dt_1} - v \right)}{\left(-\left(\frac{v}{c^2} \right) \frac{dx_1}{dt_1} + 1 \right)} \right) \frac{dt_1}{dt_2}$$

$$= \frac{\frac{d^2x_1}{dt_1^2}}{\gamma^3 \left(-\left(\frac{v}{c^2} \right) \frac{dx_1}{dt_1} + 1 \right)^3}.$$

Similarly, as will be shown in the exercises, we have

$$\frac{d^2 y_2}{dt_2^2} = \frac{1}{\gamma^2 \left(-\left(\frac{v}{c^2}\right)\frac{dx_1}{dt_1}+1\right)^2}\left(\frac{d^2 y_1}{dt_1^2}+\left(\frac{v\frac{dy_1}{dt_1}}{c^2-v\frac{dx_1}{dt_1}}\right)\frac{d^2 x_1}{dt_1^2}\right)$$

$$\frac{d^2 z_2}{dt_2^2} = \frac{1}{\gamma^2 \left(-\left(\frac{v}{c^2}\right)\frac{dx_1}{dt_1}+1\right)^2}\left(\frac{d^2 z_1}{dt_1^2}+\left(\frac{v\frac{dz_1}{dt_1}}{c^2-v\frac{dx_1}{dt_1}}\right)\frac{d^2 x_1}{dt_1^2}\right).$$

4.7. Relativistic Momentum

(While standard, here we follow Moore's *A Traveler's Guide to Spacetime: An Introduction to the Special Theory of Relativity* [43].) There are a number of measurable quantities that tell us about the world. Velocity (distance over time) and acceleration (velocity over time) are two such quantities, whose importance is readily apparent. The importance of other key quantities, such as momentum and the various types of energy, is more difficult to recognize immediately. In fact, Jennifer Coopersmith's recent *Energy, the Subtle Concept: The Discovery of Feynman's Blocks from Leibniz to Einstein* [11] gives a wonderful history of the slow, if not tortuous, history of the development of the concept of energy and, to a lesser extent, momentum. We will simply take these notions to be important. Here we will give an intuitive development of the relativistic version of momentum. As before, this should not be viewed as any type of mathematical proof. Instead, we will make a claim as to what momentum should be in relativity theory and then see, in the next chapter, that this relativistic momentum will provide us with a deeper understanding of the link between electric fields and magnetic fields.

Classically, momentum is

$$\text{momentum} = \text{mass} \times \text{velocity}.$$

Thus, classically, for a particle moving at a constant velocity of v we have

$$\text{momentum} = \text{mass} \times \frac{\text{displacement vector}}{\text{change in time}}.$$

In special relativity, though, as we have seen, the displacement vector is not natural at all. This suggests that we replace the preceding numerator with the relativistic displacement.

We need to decide now what to put into the denominator. Among all possible coordinate systems for our particle, there is one coordinate system that gives us proper time. Recall that proper time is the time coordinate in the coordinate system where the particle stays at the origin. This leads to

Definition 4.7.1. *The* relativistic momentum *p is*

$$p = \text{mass} \times \frac{\text{relativistic displacement vector}}{\text{change in proper time}}.$$

We now want formulas, since, if looked at quickly, it might appear that we are fixing a coordinate system where the particle has zero velocity and hence would always have zero momentum, which is silly.

Choose any coordinate system t', x', y', and z' in which the particle is moving at some constant $v = (v_x, v_y, v_z)$ whose speed we also denote by v. The relativistic displacement vector is

$$(c\Delta t', \Delta x', \Delta y', \Delta z').$$

In the different coordinate system that gives us the proper time (in which the particle is not moving at all), if Δt measures the change in proper time, we know from three sections ago that

$$\Delta t' = \gamma \Delta t.$$

Then for the momentum we have

$$p = (p_t, p_x, p_y, p_z)$$
$$= m \cdot \left(\frac{c\Delta t'}{\Delta t}, \frac{\Delta x'}{\Delta t}, \frac{\Delta y'}{\Delta t}, \frac{\Delta z'}{\Delta t} \right)$$
$$= m \cdot \left(\frac{c\Delta t'}{\Delta t'} \cdot \frac{\Delta t'}{\Delta t}, \frac{\Delta x'}{\Delta t'} \cdot \frac{\Delta t'}{\Delta t}, \frac{\Delta y'}{\Delta t'} \cdot \frac{\Delta t'}{\Delta t}, \frac{\Delta z'}{\Delta t'} \cdot \frac{\Delta t'}{\Delta t} \right)$$
$$= \left(\gamma mc, \gamma m \frac{\Delta x'}{\Delta t'}, \gamma m \frac{\Delta y'}{\Delta t'}, \gamma m \frac{\Delta z'}{\Delta t'} \right)$$
$$= \left(\gamma mc, \gamma m v_x, \gamma m v_y, \gamma m v_z \right).$$

Thus momentum in relativity theory is a four-vector, with spatial components the classical momentum times an extra factor of γ, and is hence

$$\gamma m v = \gamma m (v_x, v_y, v_z).$$

Our justification for this definition will lie in the work of the next chapter.

Also, though we will not justify this, the time component $p_t = \gamma mc$ of the momentum is the classical energy divided by the speed of light c. This is why in relativity theory people usually do not talk about the momentum vector but instead use the term *energy-momentum* four-vector.

4.8. Appendix: Relativistic Mass

4.8.1. Mass and Lorentz Transformations

In the last section, we gave an intuitive argument for why the momentum (in the spatial coordinates) should be $\gamma m v = \gamma m(v_x, v_y, v_z)$. Thus, in relativity theory, we want not only lengths and times to depend on reference frames but also momentums. There is another, slightly more old-fashioned approach, though. Here we stick with momentum being mass times velocity, as in classical physics, with no apparent extra factor of γ. What changes is that we no longer have the mass m as a constant but, instead, also consider the mass as depending on the reference frame.

With this approach, we start with an object that is in a reference frame in which it is not moving. Suppose it has a mass m_0, which we call the *rest mass*. In this approach, the rest mass is what, in the last section, we just called the mass. We will show that the same object moving at speed v will have mass

$$m = \gamma m_0,$$

where $\gamma = 1/\sqrt{1 - (v/c)^2}$. Then, if momentum is mass times velocity, we will have the momentum being $\gamma m_0 v$, agreeing with our work from last section.

This argument is more subtle and demands a few more underlying assumptions than those for velocity and acceleration, but it does have some advantages for justifying last section's definition for momentum. We will be closely following section 24 of [65].

In frame 1, suppose we have two identical objects moving toward each other along the x-axis, hitting each other at the origin in a head-on collision. Let Object 1 move with speed u in the x-direction and Object 2 move with speed $-u$, also in the x-direction, each with the same mass. We assume that they bounce off each other, with the first object now having speed $-u$ and the second having speed u. Such collisions are called *elastic*.

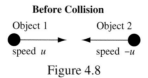

Before Collision

Object 1   Object 2

speed u speed $-u$

Figure 4.8

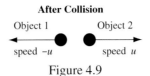

After Collision

Figure 4.9

Now assume that reference frame 2 is moving with speed $-v$ in the x-direction with respect to the lab frame. We denote the speed of Object 1 in the x-direction as u_1 and its mass as m_1, and denote the speed of Object 2 moving in the x-direction as u_2 and its mass as m_2.

Now to make some assumptions, which are a fleshing out of
Assumption I: Physics must be the same in all frames of references that move at constant velocities with respect to each other.

First, in any given reference frame, we assume that the total mass of a system cannot change (meaning that the total mass must be conserved in a given reference frame). Note that we are not saying that mass stays the same in different reference systems, just that within a fixed reference frame the total mass is a constant. In the second reference frame, let the total mass be m. Then we have

$$m_1 + m_2 = m.$$

We now set momentum to be equal to mass times velocity. (This is only for this Appendix; for most of the rest of the book, momentum will be γ times mass times velocity.) Classical physics tells us that momentum is conserved. Like mass, total momentum will be assumed to be conserved in a fixed reference frame. (Again, this is subtle, as momentum does change if we shift reference frames.)

In the second reference frame, the total momentum will be $m_1 u_1 + m_2 u_2$ before the collision. At the moment of collision, we can think of the two objects as a single object, with mass m moving with velocity v and having momentum mv. Since we are assuming that the total momentum will be conserved in reference frame 2, we have

$$m_1 u_1 + m_2 u_2 = mv.$$

We want to find formulas for the masses m_1 and m_2 in frame 2. Since we now have two equations and our two unknowns, we can solve to get

$$\frac{m_1}{m_2} = \frac{u_2 - v}{v - u_1},$$

as shown in the exercises. Note that the total mass m has dropped out.

By our work on how velocities change under Lorentz transformations, we know that

$$u_1 = \frac{u+v}{1+\dfrac{uv}{c^2}} \quad \text{and} \quad u_2 = \frac{-u+v}{1-\dfrac{uv}{c^2}}.$$

This will yield

$$\frac{m_1}{m_2} = \frac{1+\dfrac{uv}{c^2}}{1-\dfrac{uv}{c^2}},$$

as seen in the exercises.

One final equality is needed before reaching our goal of motivating the equation $m = \gamma m_0$. From the exercises, we know that $u_1 = (u+v)/(1+uv/c^2)$ implies that

$$\sqrt{1-\frac{u_1^2}{c^2}} = \frac{\sqrt{1-\dfrac{u^2}{c^2}}\sqrt{1-\dfrac{v^2}{c^2}}}{1+\dfrac{uv}{c^2}},$$

with a similar formula for $\sqrt{1-(u_2/c)^2}$.

Putting these together yields

$$\frac{m_1}{m_2} = \frac{\sqrt{1-\dfrac{u_2^2}{c^2}}}{\sqrt{1-\dfrac{u_1^2}{c^2}}}.$$

By assumption both objects are identical with mass m_0 at rest. Choose our second reference frame so that $u_2 = 0$, meaning that the second object is at rest (and hence $m_2 = m_0$) and so $u = v$. Then we get

$$\frac{m_1}{m_0} = \frac{1}{\sqrt{1-\dfrac{u_1^2}{c^2}}},$$

giving us our desired

$$m = \gamma m_0.$$

Of course, in the real world we would now have to check experimentally that momentum and mass are indeed conserved in each reference frame. As mathematicians, though, we can be content with just taking these as given. Also, as mentioned before, most people would now consider mass to be the rest mass and hence an invariant and then slightly alter the definition of momentum by multiplying through by the extra factor of γ.

4.8.2. *More General Changes in Mass*

We now want to compare the measurement of the mass of an object in different reference frames in an even more general situation.

As usual, start with reference frame 1 and assume that reference frame 2 is moving in the positive x-direction with respect to frame 1. We will have three different velocities. Let v be the velocity of frame 2 with respect to frame 1, let v_1 be the speed of an object in frame 1 moving in the x-direction, and let v_2 be the speed of the same object but now in frame 2, still moving in the x-direction. Let m_1 be the object's mass in frame 1 and m_2 be the mass in frame 2.

Our goal is to justify

$$m_2 = \left(\frac{1 - \dfrac{v_1 v}{c^2}}{\sqrt{1 - \dfrac{v^2}{c^2}}} \right) m_1$$

$$= \gamma \left(1 - \frac{v_1 v}{c^2} \right) m_1.$$

We start with the formulas for rest mass from the last section

$$m_2 = \frac{m_0}{\sqrt{1 - \dfrac{v_2^2}{c^2}}}$$

$$m_1 = \frac{m_0}{\sqrt{1 - \dfrac{v_1^2}{c^2}}}.$$

Then we get

$$m_2 = \left(\frac{\sqrt{1 - \dfrac{v_1^2}{c^2}}}{\sqrt{1 - \dfrac{v^2}{c^2}}} \right) m_1.$$

Now, we know how to relate the velocities v_1 and v_2, namely,

$$v_2 = \frac{v_1 - v}{1 - \dfrac{v_1 v}{c^2}}.$$

Plugging in the preceding, after squaring and doing a lot of algebra (which you are asked to do as an exercise), yields our desired

$$m_2 = \gamma \left(1 - \frac{v_1 v}{c^2} \right) m_1.$$

4.9. Exercises

Exercise 4.9.1. *Suppose we set up two coordinate systems. The first we will call the laboratory frame, denoting its coordinates by*

$$(x_1, y_1, z_1, t_1).$$

The second frame of reference will have coordinates denoted by

$$(x_2, y_2, z_2, t_2).$$

Suppose that this second frame is moving along at a constant speed of v in the direction parallel to the y-axis, with respect to the lab frame. Finally, suppose that the origins of both coordinate systems describe the same event. Find the Lorentz transformation that takes the (x_1, y_1, z_1, t_1) to the (x_2, y_2, z_2, t_2) coordinate system.

Exercise 4.9.2. *Suppose we set up two coordinate systems. The first we will call the laboratory frame, denoting its coordinates by*

$$(x_1, y_1, z_1, t_1).$$

Now suppose we have another frame of reference with its coordinates denoted by

$$(x_2, y_2, z_2, t_2).$$

Suppose that this new frame is moving along at a constant speed of v units/sec in the direction parallel to the vector $(1, 1, 0)$ with respect to the lab frame. Suppose that the origins of both coordinate systems describe the same event. Find the Lorentz transformation that takes the (x_1, y_1, z_1, t_1) to the (x_2, y_2, z_2, t_2) coordinate system.

Exercise 4.9.3. *Let $A \in O(3, \mathbb{R})$. Suppose that $v \in \mathbb{R}^3$ is an eigenvector of A with eigenvalue λ, meaning that $Av = \lambda v$. Show that $\lambda = \pm 1$.*

Exercise 4.9.4. *Using the previous exercise, show that if $A \in O(3, \mathbb{R})$ then A is an invertible matrix.*

Exercise 4.9.5. *Show that $A \in O(3, \mathbb{R})$ means that*

$$A^T A = I.$$

Exercise 4.9.6. *For any two 3×3 matrices, show that $(AB)^T = B^T A^T$. Use this to show that if $A, B \in O(3, \mathbb{R})$, then $AB \in O(3, \mathbb{R})$.*

Exercise 4.9.7. *Using the notation of Section 4.4.1 and using that $x_2 = \gamma x_1 - \gamma v t_1$ and $x_1 = \gamma x_2 + \gamma v t_2$ show that $t_2 = -\gamma \left(\frac{v}{c^2} \right) x_1 + \gamma t_1$.*

Exercise 4.9.8. *With the notation of Section 4.4.4, show that*

$$c^2 t_2^2 - x_2^2 - y_2^2 - z_2^2 = c^2 t_1^2 - x_1^2 - y_1^2 - z_1^2.$$

Exercise 4.9.9. *Show that if A and B are Lorentz transformations, then so is AB.*

Exercise 4.9.10. *Show that if A is a Lorentz transformation, then A must be invertible. Then show that A^{-1} is also a Lorentz transformation.*

These last two exercises show that the Lorentz transformations form a *group.*

Exercise 4.9.11. *For coordinate change*

$$x_2 = \gamma x_1 - \gamma v t_1$$
$$y_2 = y_1$$
$$z_2 = z_1$$
$$t_2 = -\gamma \left(\frac{v}{c^2}\right) x_1 + \gamma t_1,$$

where $\gamma = 1/\sqrt{1 - (\frac{v}{c})^2}$, show

$$\frac{dy_2}{dt_2} = \frac{\dfrac{dy_1}{dt_1}}{\left(-\gamma \left(\dfrac{v}{c^2}\right) \dfrac{dx_1}{dt_1} + \gamma\right)}$$

$$\frac{dz_2}{dt_2} = \frac{\dfrac{dz_1}{dt_1}}{\left(-\gamma \left(\dfrac{v}{c^2}\right) \dfrac{dx_1}{dt_1} + \gamma\right)}.$$

Exercise 4.9.12. *Using the notation from the previous exercise, show*

$$\frac{d^2x_2}{dt_2^2} = \frac{\dfrac{d^2x_1}{dt_1^2}}{\gamma^3 \left(-\left(\dfrac{v}{c^2}\right) \dfrac{dx_1}{dt_1} + 1 \right)^3}$$

$$\frac{d^2y_2}{dt_2^2} = \frac{1}{\gamma^2 \left(-\left(\dfrac{v}{c^2}\right) \dfrac{dx_1}{dt_1} + 1 \right)^2} \left(\frac{d^2y_1}{dt_1^2} + \frac{v\dfrac{dy_1}{dt_1}}{c^2 - v\dfrac{dx_1}{dt_1}} \frac{d^2x_1}{dt_1^2} \right)$$

$$\frac{d^2z_2}{dt_2^2} = \frac{1}{\gamma^2 \left(-\left(\dfrac{v}{c^2}\right) \dfrac{dx_1}{dt_1} + 1 \right)^2} \left(\frac{d^2z_1}{dt_1^2} + \frac{v\dfrac{dz_1}{dt_1}}{c^2 - v\dfrac{dx_1}{dt_1}} \frac{d^2x_1}{dt_1^2} \right).$$

Exercise 4.9.13. *Choose units such that the speed of light $c = 10$. Suppose you are moving in some coordinate frame with velocity $v = (3,5,2)$. Observe two events, the first occurring at $t_1 = 8$ and the second at $t_2 = 12$. Find the proper time between these two events.*

Exercise 4.9.14. *Choose units such that the speed of light $c = 10$. Let a particle have mass $m = 3$ and velocity $v = (4,5,2)$. Calculate its relativistic momentum.*

Exercise 4.9.15. *Using the notation from the Appendix, show that if*

$$m_1 + m_2 = m.$$

and

$$m_1 u_1 + m_2 u_2 = mv,$$

then

$$\frac{m_1}{m_2} = \frac{u_2 - v}{v - u_1}.$$

Exercise 4.9.16. *Using the notation from the Appendix, show that if*

$$\frac{m_1}{m_2} = \frac{u_2 - v}{v - u_1},$$

and

$$u_1 = \frac{u + v}{1 + \dfrac{uv}{c^2}} \quad \text{and} \quad u_2 = \frac{-u + v}{1 - \dfrac{uv}{c^2}},$$

then

$$\frac{m_1}{m_2} = \frac{1 + \dfrac{uv}{c^2}}{1 - \dfrac{uv}{c^2}},$$

Exercise 4.9.17. *Using the notation from the Appendix, show that* $u_1 = \frac{u+v}{1+uv/c^2}$
implies that

$$\sqrt{1 - \frac{u_1^2}{c^2}} = \frac{\sqrt{1 - \dfrac{u^2}{c^2}}\sqrt{1 - \dfrac{v^2}{c^2}}}{1 + \dfrac{uv}{c^2}}$$

and that $u_2 = \frac{-u+v}{1-uv/c^2}$ *implies that*

$$\sqrt{1 - \frac{u_2^2}{c^2}} = \frac{\sqrt{1 - \dfrac{u^2}{c^2}}\sqrt{1 - \dfrac{v^2}{c^2}}}{1 - \dfrac{uv}{c^2}},$$

Exercise 4.9.18. *Using the notation from the Appendix, show that*

$$m_2 = \left(\frac{1 - \dfrac{v_1 v}{c^2}}{\sqrt{1 - \dfrac{v^2}{c^2}}}\right) m_1$$

$$= \gamma \left(1 - \frac{v_1 v}{c^2}\right) m_1,$$

using that

$$m_2 = \frac{m_0}{\sqrt{1 - \dfrac{v_2^2}{c^2}}}$$

$$m_1 = \frac{m_0}{\sqrt{1 - \dfrac{v_1^2}{c^2}}}$$

$$v_2 = \frac{v_1 - v}{1 - \dfrac{v_1 v}{c^2}}.$$

5

Mechanics and Maxwell's Equations

Summary: Despite linking the electric and magnetic fields, Maxwell's equations are worthless for science if they do not lead to experimental predictions. We want to set up a formalism that allows us to make measurements. This chapter will first give an overview of Newtonian mechanics. Then we will see how the electric and magnetic fields fit into Newtonian mechanics. In the final section, we will see that the force from the electric field (Coulomb's law), together with the Special Theory of Relativity and the assumption of charge conservation, leads to magnetism.

5.1. Newton's Three Laws

The development of Newtonian mechanics is one of the highlights of humanity. Its importance to science, general culture, and our current technological world cannot be overstated. With three laws, coupled with the calculational power of calculus, much of our day-to-day world can be described. Newton, with his laws, could describe both the motions of the planets and the flight of a ball. The world suddenly became much more manageable. Newton's approach became the model for all of learning. In the 1700s and 1800s, bright young people, at least in Europe, wanted to become the Newtons of their fields by finding analogs of Newton's three laws. No one managed to become, though, the Newton of sociology.

We will state Newton's three laws and then discuss their meaning. We quote the three laws from Halliday and Resnick's *Physics* [32].

Newton's First Law: Every body persists in its state of rest or of uniform motion in a straight line unless it is compelled to change that state by forces impressed on it. (On page 75 in [32], quoting in turn Newton himself.)

Newton's Second Law:

$$\text{Force} = \text{mass} \cdot \text{acceleration}.$$

Newton's Third Law: To every action there is always an opposed equal reaction; or, the mutual actions of two bodies upon each other are always equal, and directed to contrary parts. (On page 79 in [32], quoting again Newton himself.)

Now we have to interpret what these words mean. To define terms such as "force" rigorously is difficult; we will take a much more pragmatic approach.

We start with the description of a state of an object that we are interested in. All we know is its position. We want to know how its position will change over time. We assume that there is some given Cartesian coordinate system x, y, z describing space and another coordinate t describing time. We want to find three functions $x(t)$, $y(t)$, and $z(t)$ so that the vector-valued function

$$r(t) = (x(t), y(t), z(t))$$

describes the *position* of our object for all time. Once we have $r(t)$, we can define the *velocity* $v(t)$ and the *acceleration* $a(t)$ as follows:

$$v(t) = \frac{d}{dt}r(t) = \left(\frac{dx}{dt}, \frac{dy}{dt}, \frac{dz}{dt}\right)$$

$$a(t) = \frac{d}{dt}v(t) = \left(\frac{d^2x}{dt^2}, \frac{d^2y}{dt^2}, \frac{d^2z}{dt^2}\right).$$

Though few physicists would be nervous at this point, do note that we are making assumptions about the differentiability of the component functions of the position function $r(t)$.

Now to define *force*. To some extent, Newton's three laws give a description of the mathematical possibilities for force. Newton's first law states that if there are no forces present, then the velocity must be a constant. Turning this around, we see that if the velocity is not constant, then a force must be present.

The second law

$$\text{Force} = \text{mass} \times \text{acceleration}$$

$$F = m \cdot a(t)$$

gives an explicit description of how a force affects the velocity. The right-hand side is just the second derivative of the position vector times the object's mass. It is the left-hand side that can change, depending on the physical situation we are in. For example, the only force that Newton explicitly described was gravity. Let our object have mass m in space at the point (x, y, z) in a gravitational field created by some other object of mass M centered at the origin $(0, 0, 0)$. Then the gravitational force on our first object is the vector valued function

$$F = G\frac{mM}{(x^2 + y^2 + z^2)^{3/2}}(x, y, z)$$

where G is a constant that can be determined from experiments. Note that the length of this vector, which corresponds to the magnitude of the force of gravity, is

$$|F| = G\frac{mM}{(\text{distance})^2},$$

as seen in the exercises. (We needed to put the somewhat funny looking exponent $3/2$ into the definition of the gravitational force in order to get that the magnitude of the force is proportional to 1 over the distance squared.)

In order to use the second law, we need to make a guess as to the functional form of the relevant force. Once we have that, we have a differential equation that we can attempt to solve.

The third law states, "To every action there is always an opposed and equal reaction," leading generations of students to puzzle out why there is any motion at all. (After all, why does everything not cancel out?) What Newton's third law does is allow us to understand other forces. Let us have two objects A and B interact. We say that object A acts on object B if A exerts some force on B, meaning from the first law that somehow the presence of A causes the velocity of B to change. The third law then states that B must do the same to A, but in the opposite direction. This law places strong restrictions on the nature of forces, allowing us to determine their mathematical form.

We have not explained at all the concept of mass. Also, the whole way we talked about one object exerting a force on another certainly begs a lot of questions, for example, how does one object influence another; how long does it take; or what is the mechanism? Questions such as "action at a distance" and "instantaneous transmission of forces" again become critical in relativity theory.

5.2. Forces for Electricity and Magnetism

5.2.1. $F = q(E + v \times B)$

For science to work, we must be able to make experimental predictions. We want to set up the formalism that allows us to make measurements. For now, we will be in a strictly Newtonian world. Historically, Newton's mechanics were the given and Maxwell's equations had to be justified. In fact, it strongly appears that Maxwell's equations are more basic than Newton's. But, for now, we will be Newtonian.

Maxwell's equations link the electric field with the magnetic field. But neither of these vector fields is a force and thus neither can be applied directly in Newton's second law $F = m \cdot \frac{d^2 r}{dt^2}$. We need to show how the electric field E and the magnetic field B define the force that a particle will feel. We will simply state the relationship. The "proof" lies in doing experiments.

Let a particle be described by $r(t) = (x(t), y(t), z(t))$, with velocity

$$v(t) = \frac{dr}{dt} = \left(\frac{dx}{dt}, \frac{dy}{dt}, \frac{dz}{dt} \right).$$

Suppose it has charge q, which is just a number. From experiment, the force from the electric and magnetic fields is

$$F = q(E + v \times B).$$

Thus if the particle's charge is zero, then the electric and magnetic fields will have no effect on it. If a particle's velocity is zero, then the magnetic field has no effect. The presence of the cross-product reflects what must have been initially a deep surprise, namely, that if a charged particle is moving in a given direction in the presence of a magnetic field, then the particle will suddenly accelerate at right angles to both the magnetic field and its initial direction. If the particle is initially moving in the same direction as the magnetic field, then no force is felt. We will see later in this chapter how this perpendicularity stems from the Special Theory of Relativity and the force from the electric field (namely, qE).

5.2.2. Coulomb's Law

Suppose that we have two charges, q_1 and q_2, that are a distance r apart from each other. Coulomb's law states that the magnitude of the force between them is

$$F = \text{constant} \, \frac{q_1 q_2}{r^2}.$$

This is not a mathematical theorem but an experimental fact. If r denotes the vector between the two charges, then Coulomb's law can be expressed as

$$F = \text{constant} \, \frac{q_1 q_2}{(\text{length of } r)^3} r,$$

where F denotes the force vector.

We want to use the description of the force from Coulomb's law to derive a corresponding electric field E. Fix a charge q_1 at a point in space $p_1 =$

(x_1, y_1, z_1). This charge will create an electric field, which we will now describe. Given a point $p = (x, y, z) \neq p_1$, let

$$r = (x - x_1, y - y_1, z - z_1)$$

with length

$$|r| = \sqrt{(x - x_1)^2 + (y - y_1)^2 + (z - z_1)^2}.$$

Introduce a new charge q at the point $p = (x, y, z)$. The electric field created by the charge q_1 will be

$$E(x, y, z) = \frac{q_1}{|r|^3} r = \left(\frac{q_1(x - x_1)}{|r|^3}, \frac{q_1(y - y_1)}{|r|^3}, \frac{q_1(z - z_1)}{|r|^3} \right),$$

since this electric field will indeed give us that $F = qE$.

5.3. Force and Special Relativity

5.3.1. The Special Relativistic Force

In classical mechanics, we have the force as

$$F = ma = m\frac{dv}{dt} = \frac{d(mv)}{dt} = \frac{d(\text{momentum})}{dt}.$$

Here we are using the classical mechanics definition for momentum, namely, that momentum equals mass times velocity. Last chapter, though, we gave an intuitive argument for why in relativity theory the momentum vector $p = (p_x, p_y, p_z)^1$ should actually be

$$p = (p_x, p_y, p_z)$$
$$= \gamma m(v_x, v_y, v_z)$$
$$= \gamma mv,$$

where $\gamma = 1/\sqrt{1 - (v/c)^2}$ with v being the speed. This leads to the question of the correct definition for force in special relativity: Is it $F = ma$ or $F = \frac{d(\gamma mv)}{dt}$? We treat this as an empirical question, allowing nature to dictate that the force must be

$$F = \frac{d(\gamma mv)}{dt}$$

[1] A comment on notation. Earlier we used the notation p_x to refer to $\partial p/\partial x$, and so forth. In this section, though, p_x refers not to the partial derivative but to the x-component of the vector p. Earlier we denoted this x-component by p_1, which we are not doing here, since in a moment we will need to use the numbered subscript to refer to different frames of reference. We will follow the same notation convention for the velocity $= (v_x, v_y, v_z)$, acceleration $a = (a_x, a_y, a_z)$, and force $F = (F_x, F_y, F_z)$.

in special relativity. (There are other more mathematical approaches.)
Denote the force as

$$F = (F_x, F_y, F_z),$$

the velocity as

$$v = (v_x, v_y, v_z),$$

and the acceleration as

$$a = (a_x, a_y, a_z).$$

Then we get

$$
\begin{aligned}
F &= (F_x, F_y, F_z) \\
&= \frac{d}{dt}(\gamma m v) \\
&= \gamma m \frac{d(v)}{dt} + m v \frac{d(\gamma)}{dt} \\
&= \gamma m (a_x, a_y, a_z) + \frac{d(\gamma)}{dt}(m v_x, m v_y, m v_z) \\
&= \gamma m a + \frac{d(\gamma)}{dt} m v.
\end{aligned}
$$

In special relativity, it is no longer the case that the force vector points in the
same direction as the acceleration vector.

5.3.2. Force and Lorentz Transformations

We now want to see how forces change under Lorentz transformations. We
could derive the equations, using the same techniques as used previously
to find how velocity, acceleration, and momentum change under Lorentz
transformations. This derivation is, though, quite algebraically complicated.
Thus we will just write down the resulting coordinate changes.

Let reference frame 1 denote the lab frame. In this frame of reference,
denote the object's velocity by

$$v_1 = (v_{1x}, v_{1y}, v_{1z})$$

and its force by

$$F_1 = (F_{1x}, F_{1y}, F_{1z}).$$

Let reference frame 2 move at constant speed v in the direction $(1,0,0)$ with respect to frame 1, with Lorentz transform

$$x_2 = \gamma x_1 - \gamma v t_1$$

$$y_2 = y_1$$

$$z_2 = z_1$$

$$t_2 = -\gamma \left(\frac{v}{c^2}\right) x_1 + \gamma t_1,$$

where $\gamma = 1/\sqrt{1 - (v/c)^2}$. Let the force in reference frame 2 be denoted by $F_2 = (F_{2x}, F_{2y}, F_{2z})$, and let its velocity be denoted by $v_1 = (v_{2x}, v_{2y}, v_{2z})$.
Then, after a lot of calculations, we get

$$F_{1x} = F_{2x} + \frac{v}{c^2 + v_{2x}v}(v_{2y}F_{2y} + v_{2z}F_{2z})$$

$$F_{1y} = \frac{F_{2y}}{\gamma\left(1 + \left(\frac{v_{2x}v}{c^2}\right)\right)}$$

$$F_{1z} = \frac{F_{2z}}{\gamma\left(1 + \left(\frac{v_{2x}v}{c^2}\right)\right)}.$$

5.4. Coulomb + Special Relativity
+ Charge Conservation = Magnetism

(While standard, this follows from section 6.1 in Lorrain and Corson's *Electromagnetic Fields and Waves* [37].)

Maxwell's equations use cross-products. This reflects the physical fact that if you have current flowing through a wire in the xz-plane running along the x-axis and then move a charge (in the yz-plane) along the y-axis toward the wire, suddenly the charge will feel a force perpendicular to the yz-plane. The charge wants to jump out of the plane. This is easy to observe experimentally. On a personal note, I still remember my uneasiness upon learning this strange fact. Thus it came as both a relief and a surprise when I learned in college how special relativity gives a compelling reason for the occurrence of this "perpendicularness."

Here is another way of interpreting the "oomph" behind this section. Maxwell's equations explain how electricity and magnetism are linked but, in their initial form, give no clue as to why they are related. Special relativity will answer this "why" question by showing that our observation of a magnetic force is actually our observation of the electric force in a moving frame of reference.

We will show in this section how

Coulomb's Law + Special Relativity + Charge Invariance

will yield

Magnetism.

We make the (experimentally verifiable) assumption that the charge of an object is independent of frame of reference. This means that the total charge does not change in a given reference frame (i.e., charge is conserved) but also that the charge is the same in every reference frame (i.e, charge is an invariant). This is in marked contrast to quantities such as length and time.

Now to see how magnetism must occur under the assumptions of Coulomb's law, special relativity, and charge invariance. We start with two charges, q_1 and q_2, and will consider two reference frames 1 and 2.

In reference frame 2, suppose that the two charges have coordinates $q_1 = (0,0,0)$ and $q_2 = (x_2, y_2, 0)$, as in Figure 5.1.

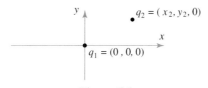

Figure 5.1

We assume in frame 2 that the charges are not moving. Then by Coulomb's law, the force between the two charges is

$$F_2 = (F_{2x}, F_{2y}, F_{2z})$$

$$= \left(\frac{q_1 q_2}{(x_2^2 + y_2^2)^{3/2}} x_2, \frac{q_1 q_2}{(x_2^2 + y_2^2)^{3/2}} y_2, 0 \right).$$

Let reference frame 2 move at constant speed v in the direction $(1,0,0)$ with respect to frame 1, with Lorentz transform

$$x_2 = \gamma x_1 - \gamma v t_1$$

$$y_2 = y_1$$

$$z_2 = z_1$$

$$t_2 = -\gamma \left(\frac{v}{c^2} \right) x_1 + \gamma t_1,$$

where $\gamma = 1/\sqrt{1 - (v/c)^2}$. Since the charges are not moving in frame 2, we have

$$v_2 = (v_{2x}, v_{2y}, v_{2z}) = (0, 0, 0).$$

Then, from the previous section, we have

$$F_{1x} = F_{2x} + \frac{v}{c^2 + v_{2x}v}(v_{2y}F_{2y} + v_{2z}F_{2z})$$

$$= F_{2x}$$

$$= \frac{q_1 q_2}{(x_2^2 + y_2^2)^{3/2}} x_2$$

$$F_{1y} = \frac{F_{2y}}{\gamma\left(1 + \left(\frac{v_{2x}v}{c^2}\right)\right)}$$

$$= \frac{1}{\gamma} F_{2y}$$

$$= \frac{q_1 q_2}{\gamma(x_2^2 + y_2^2)^{3/2}} y_2$$

$$F_{1z} = \frac{F_{2z}}{\gamma\left(1 + \left(\frac{v_{2x}v}{c^2}\right)\right)}$$

$$= 0.$$

We now put F_{1x} and F_{1y} in terms of the coordinates x_1 and y_1, yielding, at $t_1 = 0$,[2]

$$F_{1x} = \frac{\gamma q_1 q_2 x_1}{(\gamma^2 x_1^2 + y_1^2)^{3/2}}$$

$$F_{1y} = \frac{q_1 q_2 y_1}{\gamma(\gamma^2 x_1^2 + y_1^2)^{3/2}}$$

$$= \frac{\gamma q_1 q_2 y_1}{(\gamma^2 x_1^2 + y_1^2)^{3/2}}\left(1 - \left(\frac{v}{c}\right)^2\right).$$

Then the force F_1 is

$$\frac{\gamma q_1 q_2}{(\gamma^2 x_1^2 + y_1^2)^{3/2}}(x_1, y_1, 0) - \frac{\gamma q_1 q_2 v^2 y_1}{c^2(\gamma^2 x_1^2 + y_1^2)^{3/2}}(0, 1, 0),$$

---

[2] Note we are using that:

$$\frac{1}{\gamma} = \frac{\gamma}{\gamma^2}$$

$$= \gamma\left(1 - \left(\frac{v}{c}\right)^2\right).$$

which equals

$$q_2 \left(\frac{\gamma q_1}{(\gamma^2 x_1^2 + y_1^2)^{3/2}}(x_1, y_1, 0) + v(1, 0, 0) \times \left(\frac{\gamma q_1 v y_1}{c^2 (\gamma^2 x_1^2 + y_1^2)^{3/2}} \right)(0, 0, 1) \right).$$

The

$$q_2 \frac{\gamma q_1}{(\gamma^2 x_1^2 + y_1^2)^{3/2}}(x_1, y_1, 0)$$

part of the force is just the contribution, in frame 1, from Coulomb's law. But there is the additional term, pointing in the $(0, 0, 1)$ direction. It is this second term that is captured by magnetism.

5.5. Exercises

Exercise 5.5.1. *Let*

$$F = G \frac{mM}{(x^2 + y^2 + z^2)^{3/2}}(x, y, z)$$

where G, m, and M are constants. Show that the magnitude of F is

$$|F| = G \frac{mM}{(\text{distance})^2}.$$

The next few problems discuss how to solve second order ordinary differential equations, which naturally occur when trying to solve

$$F = ma = m \frac{d^2 r}{dt^2}.$$

Exercise 5.5.2. *Suppose we have an object on a spring with mass $m = 1$ that can only move along the x-axis. We know that the force F on a spring is*

$$F(x, t) = -k^2 x,$$

where k is a constant. Let $x(t)$ describe the path of the object. Suppose that we know

$$x(0) = 1$$
$$\frac{dx}{dt}(0) = 1.$$

Using Newton's second law, show that

$$x(t) = \cos(kt) + \frac{1}{k} \sin(kt)$$

is a possible solution.

Exercise 5.5.3. *Let $y_1(t)$ and $y_2(t)$ be solutions to the ordinary differential equation*

$$\frac{d^2 y(t)}{dt^2} + a\frac{dy(t)}{dt} + by = 0,$$

where a and b are constants. Show that

$$\alpha y_1(t) + \beta y_2(t)$$

is also a solution, for any constants α, β.

Exercise 5.5.4. *Consider the ordinary differential equation*

$$\frac{d^2 y(t)}{dt^2} + a\frac{dy(t)}{dt} + by = 0,$$

for two constants a and b. Show that if

$$y(t) = e^{\alpha t}$$

is a solution, then α must be a root of the second-degree polynomial

$$x^2 + ax + b.$$

If $a^2 - 4b \neq 0$, find two linearly independent solutions to this differential equation. (Recall that functions $f(t)$ and $g(t)$ are linearly independent when the only way for there to be constants λ and μ such that

$$\lambda f(t) = \mu g(t)$$

is if

$$\lambda = \mu = 0.)$$

Exercise 5.5.5. *Using Taylor series, show that for all $t \in \mathbb{R}$,*

$$e^{it} = \cos(t) + i\sin(t)$$

$$e^{-it} = \cos(t) - i\sin(t).$$

Exercise 5.5.6. *Find two linearly independent solutions, involving exponential functions, to*

$$\frac{d^2 y(t)}{dt^2} + k^2 y = 0,$$

via finding the roots to

$$t^2 + k^2 = 0.$$

Use these solutions to show that

$$\alpha \cos(kt) + \beta \sin(kt)$$

is also a solution, for any constants α, β.

The next series of exercises discusses a special type of force, called conservative. Recall that a path in $\mathbb{R}^3$ is given by

$$r(t) = (x(t), y(t), z(t)).$$

A tangent vector to this path at a point $r(t_0)$ is

$$\frac{dr(t_0)}{dt} = \left(\frac{dx(t_0)}{dt}, \frac{dy(t_0)}{dt}, \frac{dz(t_0)}{dt} \right).$$

Let

$$F = (F_1(x,y,z), F_2(x,y,z), F_3(x,y,z)).$$

Then we have, for a curve σ defined by $r(t) = (x(t), y(t), z(t))$ with $a \le t \le b$,

$$\int_\sigma F \cdot dr(t) = \int_a^b F_1(x(t), y(t), z(t)) \frac{dx}{dt} dt$$

$$+ \int_a^b F_2(x(t), y(t), z(t)) \frac{dy}{dt} dt$$

$$+ \int_a^b F_3(x(t), y(t), z(t)) \frac{dz}{dt} dt.$$

Exercise 5.5.7. *Let* σ *be defined by* $r(t) = (x(t), y(t), z(t))$ *with* $a \le t \le b$. *Suppose*

$$r(a) = p = (p_1, p_2, p_3)$$
$$r(b) = q = (q_1, q_2, q_3).$$

Finally, for a function $f(x,y,z)$, *set*

$$F = \nabla(f).$$

Then show that

$$\int_\sigma F \cdot dr(t) = f(q_1, q_2, q_3) - f(p_1, p_2, p_3).$$

Any vector field that is defined via $F = \nabla(f)$, for a function $f(x,y,z)$, is said to be *conservative*.

Exercise 5.5.8. *Let* F *be a conservative vector field. Let* σ_1 *be defined by* $r_1(t) = (x_1(t), y_1(t), z_1(t))$ *with* $a \le t \le b$ *and let* σ_2 *be defined by* $r_2(t) =$

$(x_2(t), y_2(t), z_1(t))$ with $a \leq t \leq b$ such that

$$r_1(a) = r_2(a), \quad r_1(b) = r_2(b).$$

Then show that

$$\int_{\sigma_1} F \cdot dr_1(t) = \int_{\sigma_2} F \cdot dr_2(t).$$

Exercise 5.5.9. *Let F be a conservative vector field. Let σ be defined by $r(t) = (x(t), y(t), z(t))$ with $a \leq t \leq b$ such that*

$$r(a) = r(b).$$

Show that

$$\int_{\sigma} F \cdot dr(t) = 0.$$

Exercise 5.5.10. *Let F be conservative. Show that*

$$\nabla \times F = (0, 0, 0).$$

Exercise 5.5.11. *By considering the function*

$$f(x, y, z) = -G \frac{Mm}{\sqrt{x^2 + y^2 + z^2}},$$

show that the gravitational force

$$F(x, y, z) = G \frac{Mm}{(x^2 + y^2 + z^2)^{3/2}} (x, y, z)$$

is conservative.

Exercise 5.5.12. *Let reference frame 1 denote the lab frame. In this frame of reference, denote the object's velocity by*

$$v_1 = (v_{1x}, v_{1y}, v_{1z})$$

and its force by

$$F_1 = (F_{1x}, F_{1y}, F_{1z}).$$

Let reference frame 2 move at constant speed v in the direction $(0, 1, 0)$ with respect to frame 1, with Lorentz transformation

$$x_2 = x_1$$

$$y_2 = \gamma y_1 - \gamma v t_1$$

$$z_2 = z_1$$

$$t_2 = -\gamma \left(\frac{v}{c^2} \right) x_1 + \gamma t_1,$$

where $\gamma = 1/\sqrt{1-(v/c)^2}$. *Let the force in reference frame 2 be denoted by* $F_2 = (F_{2x}, F_{2y}, F_{2z})$ *and let its velocity be denoted by* $v_2 = (v_{2x}, v_{2y}, v_{2z})$. *Write* F_1 *in terms of the components of the force* F_2 *and the velocity* v_2.

Exercise 5.5.13. *Use the notation of Section 5.4, but now assume that the second charge* q_2 *is at the point* $(x_2, 0, z_2)$. *Find the force* F_2.

Exercise 5.5.14. *Using the notation of the previous problem, calculate the force* F_1 *and identify what should be called the magnetic component.*

Exercise 5.5.15. *Use the notation of Section 5.4, but now assume that the second charge* q_2 *is at the point* $(x_2, 0, 0)$. *Find the force* F_2. *Then calculate the force* F_1 *and identify what should be called the magnetic component.*

Exercise 5.5.16. *Use the notation of Section 5.4, but now assume that the second charge* q_2 *is at the point* $(0, y_2, 0)$. *Find the force* F_2. *Then calculate the force* F_1 *and identify what should be called the magnetic component.*

6

Mechanics, Lagrangians, and the Calculus of Variations

Summary: The goal of this chapter is to recast all of mechanics into the problem of finding critical points over a space of functions of an integral (the integral of what will be called the Lagrangian). This process is called the calculus of variations. Using the tools of the calculus of variations, we will show that $F = ma$ is equivalent to finding critical points of the integral of the Lagrangian. This approach is what most naturally generalizes to deal with forces beyond electricity and magnetism and will be the natural approach that we will later take when we quantize Maxwell's equations.

6.1. Overview of Lagrangians and Mechanics

Newton's second law states that

$$\text{Force} = \text{mass} \times \text{acceleration}.$$

As discussed in the last chapter, this means that if we want to predict the path of a particle $(x(t), y(t), z(t))$ with mass m, given a force

$$(F_1(x, y, z, t), F_2(x, y, z, t), F_3(x, y, z, t)),$$

we have to solve the differential equations

$$F_1(x, y, z, t) = m \frac{d^2 x}{dt^2}$$

$$F_2(x, y, z, t) = m \frac{d^2 y}{dt^2}$$

$$F_3(x, y, z, t) = m \frac{d^2 z}{dt^2}.$$

In this chapter, we will see how this system of differential equations can be recast into a system of integrals. The motion of the underlying object will no longer be modeled as a function that solves a system of differential equations but now as a function that is a critical point of a system of integrals. This system of integrals is called the *action*. The function that is integrated is called the *Lagrangian*. The process of finding these critical paths is called the *calculus of variations*.

Newtonian mechanics is only one theory of mechanics, which, while working quite well for most day-to-day calculations, is not truly accurate, as we saw in the need for the Special Theory of Relativity. In fact, as we will see in a few chapters, we will eventually need a completely different theory of mechanics: quantum mechanics. Moving even beyond quantum mechanics, when trying to understand the nature of elementary particles, Newtonian mechanics breaks down even further. For these new theories of mechanics, the Lagrangian approach for describing the time evolution of the state of a particle is what can be most easily generalized. We will see that different theories of mechanics are distinguished by finding critical points of the integrals with different Lagrangians. The physics is captured by the Lagrangian.

In this chapter, we will develop the calculus of variations and then see how Newton's second law can be recast into this new language.

6.2. Calculus of Variations

The differential equations of mechanics have natural analogs in the calculus of variations. The formulations in terms of the calculus of variations are in fact the way to understand the underlying mechanics involving the weak and strong forces. In the next subsection, we will set up the basic framework for the calculus of variations. Then, we will derive the Euler-Lagrange equations, which are the associated system of differential equations.

6.2.1. Basic Framework

One of the basic problems that differential calculus of one variable solves is that of finding maximums and minimums for functions. For a function to have a max or a min at a point p for some reasonable function $f(x)$, it must be the case that its derivative at p must be zero:

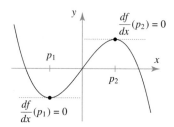

Figure 6.1

Calculus gives us a procedure for finding the points where a function has an extreme point.

Remember that one subtlety from calculus in finding local minimums and maximums is that there can be points where the derivative is zero that are neither minimums nor maximums. For example, at $x = 0$, the function $f(x) = x^3$ has derivative zero, despite the fact that the origin is not a minimum or a maximum. Back in calculus, points for which the derivative is zero are called *critical* or *extreme* points. All we know is that the local minimums and local maximums are contained in the extreme points.

The calculus of variations provides us with a mechanism not for finding isolated points that are the extreme values of a function but for finding curves along which an integral has a max or a min. As with usual calculus, these are called *critical* or *extreme* curves. The curves that minimize or maximize the integrals will be among the extreme curves. For ease of exposition, we will be finding the curves that minimize the integrals.

Consider $\mathbb{R}^2$ with coordinates (t, x) and some reasonable function $f(t, x, y)$ of three variables. (By reasonable, we mean that the following integrals and derivatives all exist.) Fix two points (t_0, x_0) and (t_1, x_1). We want to find a function $x(t)$ that makes the integral

$$\int_{t_0}^{t_1} f(t, x(t), x'(t)) \, dt$$

as small as possible. (Just to be clear, $x'(t)$ is the derivative of the function $x(t)$.)

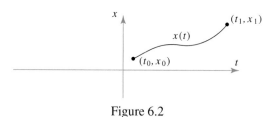

Figure 6.2

For example, let $f(t,x,y) = x^2 + y^2$ and let

$$(t_0, x_0, y_0) = (1,0,0) \text{ and } (t_1, x_1, y_1) = (3,0,0).$$

For any function $x(t)$, we have

$$\int_1^3 (x(t)^2 + x'(t)^2) \, dt \geq 0.$$

This will be minimal precisely when $x(t) = 0$. Of course, for most examples we cannot find an answer by just looking at the integral.

6.2.2. Euler-Lagrange Equations

We would like to be able to find some reasonable description for the solution curve $x(t)$ to the calculus of variations problem of minimizing

$$\int_{t_0}^{t_1} f(t, x(t), x'(t)) \, dt.$$

We reduce this to standard calculus, namely, to taking a derivative of a new one-variable function. At a key stage, we will be using both integration by parts and the multi-variable chain rule.

Start with the assumption that we somehow already know a solution, which we call $x(t)$. In particular we assume we know the initial conditions

$$x(t_0) = x_0, \ x(t_1) = x_1.$$

We want to look at paths that are slightly perturbed from the true (though for now unknown) minimal path $x(t)$. We do this by considering a function $\eta(t)$, subject only to the conditions

$$\eta(t_0) = 0, \ \eta(t_1) = 0.$$

The perturbed function will be

$$x_\epsilon(t) = x(t) + \epsilon \eta(t).$$

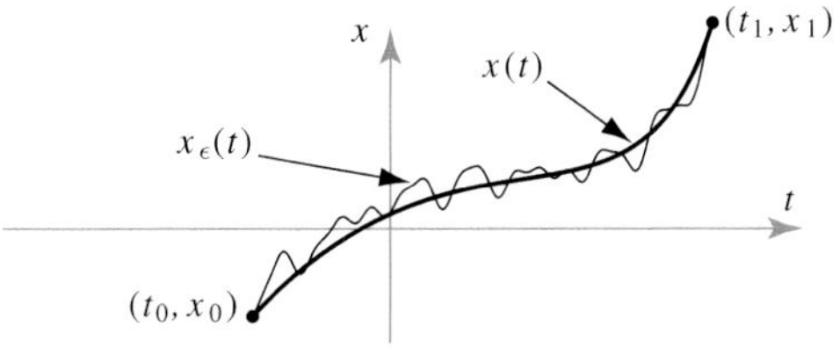

Figure 6.3

Here we of course are thinking of ϵ as some small number. The boundary conditions on $\eta(t)$ ensure that $x_\epsilon(t)$ and $x(t)$ agree at the endpoints t_0 and t_1. By assumption on the function $x(t)$, we know that for all ϵ

$$\int_{t_0}^{t_1} f(t,x(t),x'(t))\,dt \le \int_{t_0}^{t_1} f(t,x_\epsilon(t),x_\epsilon'(t))\,dt.$$

Define a new function

$$F(\epsilon) = \int_{t_0}^{t_1} f(t,x_\epsilon(t),x_\epsilon'(t))\,dt.$$

This function is a function of the real variable ϵ, which has a minimum at $\epsilon = 0$, since $x_0(t) = x(t)$. Thus its derivative must be zero at $\epsilon = 0$

$$\frac{dF}{d\epsilon}(0) = 0,$$

no matter what function η we choose. The strategy now is to differentiate $F(\epsilon)$ and to find needed conditions on the function $x(t)$. We have

$$0 = \frac{dF}{d\epsilon}(0) = \frac{d}{d\epsilon}\int_{t_0}^{t_1} f(t,x_\epsilon(t),x_\epsilon'(t))\,dt\Big|_{\epsilon=0}$$

$$= \int_{t_0}^{t_1} \frac{d}{d\epsilon} f(t,x_\epsilon(t),x_\epsilon'(t))\,dt\Big|_{\epsilon=0}.$$

Note that we have not justified our pulling the derivative inside the integral. (This can be made rigorous, as discussed in [44].) Now, by the multi-variable chain rule, we have

$$\frac{d}{d\epsilon} f(t,x_\epsilon(t),x_\epsilon'(t)) = \frac{\partial}{\partial t} f(t,x_\epsilon(t),x_\epsilon'(t))\frac{dt}{d\epsilon}$$

$$+ \frac{\partial}{\partial x} f(t,x_\epsilon(t),x_\epsilon'(t))\frac{dx_\epsilon}{d\epsilon}$$

$$+ \frac{\partial}{\partial x'} f(t,x_\epsilon(t),x_\epsilon'(t))\frac{dx_\epsilon'}{d\epsilon}$$

$$= \frac{\partial}{\partial x} f(t,x_\epsilon(t),x_\epsilon'(t))\eta(t) + \frac{\partial}{\partial x'} f(t,x_\epsilon(t),x_\epsilon'(t))\eta'(t).$$

As is usual with the multi-variable chain rule, the notation begins to become cumbersome. Here the symbols x and x' are playing the roles of both dependent variables (as functions of t and ϵ) when we differentiate and independent variables (when we write $\frac{\partial}{\partial x} f$ and $\frac{\partial}{\partial x'} f$). We are also using the fact that t is independent of the variable ϵ, meaning that $\frac{dt}{d\epsilon} = 0$.

Thus we have

$$\frac{d}{d\epsilon} f(t, x_\epsilon(t), x_\epsilon'(t)) = \frac{\partial}{\partial x} f(t, x_\epsilon(t), x_\epsilon'(t))\eta(t) + \frac{\partial}{\partial x'} f(t, x_\epsilon(t), x_\epsilon'(t))\eta'(t).$$

At $\epsilon = 0$, we get

$$0 = \int_{t_0}^{t_1} \frac{\partial}{\partial x} f(t, x(t), x'(t))\eta(t)dt + \int_{t_0}^{t_1} \frac{\partial}{\partial x'} f(t, x(t), x'(t))\eta'(t)\, dt.$$

We briefly concentrate on the term $\int_{t_0}^{t_1} \frac{\partial}{\partial x'} f(t, x(t), x'(t))\eta'(t)\, dt$. It is here that we use integration by parts (i.e., $\int u \cdot dv = u \cdot v - \int v \cdot du$). Let

$$u = \frac{\partial}{\partial x'} f(t, x(t), x'(t))$$

$$\frac{dv}{dt} = \eta'(t).$$

Using that $\eta(t_0) = 0$ and $\eta(t_1) = 0$, we get that

$$\int_{t_0}^{t_1} \frac{\partial}{\partial x'} f(t, x(t), x'(t))\eta'(t)\, dt = -\int_{t_0}^{t_1} \eta(t)\frac{d}{dt}\frac{\partial}{\partial x'} f(t, x(t), x'(t))\, dt.$$

Thus we have the function η appearing in both of the integral terms. Combining both into one integral, we get

$$0 = \int_{t_0}^{t_1} \eta(t)\left[\frac{\partial}{\partial x} f(t, x(t), x'(t)) - \frac{d}{dt}\frac{\partial}{\partial x'} f(t, x(t), x'(t))\right] dt.$$

Now to use that the function $\eta(t)$ can be any perturbation. It can be any reasonable function as long as it satisfies the boundary conditions $\eta(t_0) = 0$ and $\eta(t_1) = 0$. But no matter what function $\eta(t)$ we choose, the preceding integral must be zero. The only way that this can happen is if our solution curve $x(t)$ satisfies

$$\frac{\partial}{\partial x} f(t, x(t), x'(t)) - \frac{d}{dt}\frac{\partial}{\partial x'} f(t, x(t), x'(t)) = 0.$$

This is the Euler-Lagrange equation. Though the preceding was derived for when $x(t)$ was a minimizing curve, the same argument works for when $x(t)$ is maximizing or, in fact, when $x(t)$ is any critical value. Thus we have

Theorem 6.2.1. *A function $x(t)$ is a critical solution to the integral equation*

$$\int_{t_0}^{t_1} f(t, x(t), x'(t))\, dt$$

if and only if it satisfies the Euler-Lagrange equation

$$\frac{\partial}{\partial x} f(t,x(t),x'(t)) - \frac{d}{dt}\frac{\partial}{\partial x'} f(t,x(t),x'(t)) = 0.$$

Of course, we should now check this by doing an example whose answer we already know. In the (t,x)-plane, let us prove that the Euler-Lagrange equation will show us that the shortest distance between two points is a straight line. If the Euler-Lagrange equations did not show this, we would know we had made a mistake in the previous derivation.

Fix two points (t_0,x_0) and (t_1,x_1). Given any function $x(t)$ with

$$x(t_0) = x_0 \text{ and } x(t_1) = x_1,$$

we know that the arclength is given by the integral

$$\int_{t_0}^{t_1} \sqrt{1+(x')^2}\, dt.$$

In the language we used earlier, we have

$$f(t,x,x') = \sqrt{1+(x')^2}.$$

Note that this function does not depend on the variable x. The Euler-Lagrange equation gives us that

$$0 = \frac{\partial}{\partial x} f(t,x(t),x'(t)) - \frac{d}{dt}\frac{\partial}{\partial x'} f(t,x(t),x'(t))$$

$$= \frac{\partial}{\partial x}\sqrt{1+(x')^2} - \frac{d}{dt}\frac{\partial}{\partial x'}\sqrt{1+(x')^2}$$

$$= -\frac{d}{dt}\frac{\partial}{\partial x'}\sqrt{1+(x')^2}.$$

Since the derivative with respect to t of the function $\frac{\partial}{\partial x'}\sqrt{1+(x')^2}$ is zero, it must itself be equal to a constant, say c. Thus

$$c = \frac{\partial}{\partial x'}\sqrt{1+(x')^2}$$

$$= \frac{x'}{\sqrt{1+(x')^2}}.$$

Then

$$c\sqrt{1+(x')^2} = x'.$$

Squaring both sides and then solving gives us

$$(x')^2 = \frac{c^2}{1-c^2}.$$

Taking square roots, we see that the derivative of the function $x(t)$ must be equal to a constant:

$$x'(t) = \text{constant}.$$

Thus we indeed get that

$$x(t) = a + bt,$$

a straight line. (The constants a and b can be found from the initial conditions.)

6.2.3. More Generalized Calculus of Variations Problems

Our attempt to find an extremal function $x(t)$ for an integral

$$\int L(t, x(t), x'(t)) \, dt$$

is only one type of a calculus of variations style problem. In general, we are trying to find a function or a collection of functions that are critical points for some sort of integral. Also, in general, we can show that these critical points must satisfy some sort of collection of differential equations, all of which are called Euler-Lagrange equations. (Again, though we are calling these "critical points," they are actually functions.) The method for deriving the Euler-Lagrange equations for a given calculus of variations problem, though, is almost always analogous to what we did in the previous section: assume we have a critical point, perturb it a bit, and then reduce the original problem to finding a critical point of a one-variable function.

In this section, we just state various Euler-Lagrange equations, leaving the proofs to the exercises.

Theorem 6.2.2. *An extremum* $(x_1(t), \ldots, x_n(t))$ *to*

$$\int_{t_0}^{t_1} L(t, x_1(t), x_1'(t), \ldots, x_n(t), x_n'(t)) \, dt,$$

subject to the initial conditions

$$x(t_0) = a_1, \ldots, x_n(t_0) = a_n$$

and

$$x(t_1) = b_1, \ldots, x_n(t_1) = b_n$$

will satisfy

$$\frac{\partial}{\partial x_1}L(t,x(t),x'(t),y(t),y'(t)) - \frac{d}{dt}\frac{\partial}{\partial x_1'}L(t,x(t),x'(t),y(t),y'(t)) = 0$$

$$\vdots$$

$$\frac{\partial}{\partial x_n}L(t,x(t),x'(t),y(t),y'(t)) - \frac{d}{dt}\frac{\partial}{\partial x_n'}L(t,x(t),x'(t),y(t),y'(t)) = 0.$$

Here we are trying to find a curve $(x_1(t),\ldots,x_n(t))$ that is a critical point for some integral.

The next type of calculus of variations problem is finding surfaces $x(s,t)$ that minimize some integral.

Theorem 6.2.3. *Let R be a rectangle in the (s,t)-plane, with boundary $\partial(R)$. If a function $(x(s,t))$ minimizes the integral*

$$\int_R L\left(s,t,x(s,t),\frac{\partial x}{\partial s},\frac{\partial x}{\partial t}\right)dsdt,$$

then $x(s,t)$ will satisfy

$$\frac{\partial L}{\partial x} - \frac{\partial}{\partial s}\frac{\partial L}{\partial x_s} - \frac{\partial}{\partial t}\frac{\partial L}{\partial x_t} = 0.$$

(Here we are using the notation

$$x_s = \frac{\partial x}{\partial s}$$

$$x_t = \frac{\partial x}{\partial t}.)$$

And of course we can create new calculus of variations problems by adding more dependent and independent variables.

6.3. A Lagrangian Approach to Newtonian Mechanics

Newton's second law states that $F(x,y,z,t) = m \cdot a(x,y,z,t)$. We want to replace this differential equation with an integral equation from the calculus of variations. (At this point, it should not be clear that there is any advantage to this approach.) We have a vector-valued function $r(t) = (x(t),y(t),z(t))$ that describes a particle's position. Our goal is still explicitly to find this function $r(t)$. We still define velocity $v(t)$ as the time derivative of $r(t)$:

$$v(t) = \frac{dr}{dt} = \left(\frac{dx}{dt},\frac{dy}{dt},\frac{dz}{dt}\right).$$

Likewise, the acceleration is still the second derivative of the position function $r(t)$:

$$a(t) = \frac{d^2 r}{dt^2} = \left(\frac{d^2 x}{dt^2}, \frac{d^2 y}{dt^2}, \frac{d^2 z}{dt^2} \right).$$

We still want to see how the acceleration is related to the vector field describing force:

$$F(x, y, z, t) = (F_1(x, y, z, t), F_2(x, y, z, t), F_3(x, y, z, t)).$$

Now we must make an additional assumption on this vector field F, namely, that F is *conservative* or *path-independent*. These terms mean the same thing and agree with the terms from vector calculus. Recall that a vector field is conservative if there exists a function $U(x, y, z, t)$ such that

$$-\frac{\partial U}{\partial x} = F_1(x, y, z, t)$$

$$-\frac{\partial U}{\partial y} = F_2(x, y, z, t)$$

$$-\frac{\partial U}{\partial z} = F_3(x, y, z, t).$$

This can also be written using the gradient operator $\nabla = \left(\frac{\partial}{\partial x}, \frac{\partial}{\partial y}, \frac{\partial}{\partial z} \right)$ as

$$-\nabla U = F.$$

The function U is usually called the *potential* or, in physics if F is a force, the *potential energy*. The existence of such a function U is equivalent to saying that path integrals of F depend only on end points. (This is deep and important, but standard, in vector calculus.) Thus we assume that for any differentiable path

$$\sigma(t) = (x(t), y(t), z(t))$$

with starting point

$$\sigma(0) = (x_0, y_0, z_0)$$

and endpoint

$$\sigma(1) = (x_1, y_1, z_1),$$

the value of the integral

$$\int_\sigma F \cdot d\sigma$$

depends only on the points (x_0, y_0, z_0) and (x_1, y_1, z_1).

The potential function U and the path integral are linked by the following:

$$\int_\sigma F \cdot d\sigma = U(x_1, y_1, z_1, t) - U(x_0, y_0, z_0, t).$$

Note that we are leaving the time t fixed throughout these calculations. In many situations, such as gravitational fields, the force F is actually independent of the time t.

We need a few more definitions, whose justification will be that they will yield a calculus of variations approach to Newton's second law. Define *kinetic energy* to be

$$T = \frac{1}{2}m \cdot (\text{length of } v)^2$$

$$= \frac{1}{2}m \left(\left(\frac{dx}{dt} \right)^2 + \left(\frac{dy}{dt} \right)^2 + \left(\frac{dz}{dt} \right)^2 \right).$$

Note that this is the definition of the kinetic energy. (Using the term "energy" does help physicists' intuition. It appears that physicists have intuitions about these types of "energies." Certainly in popular culture people bandy about the word "energy" all the time. For us, the preceding ones are the definitions of energy, no more and no less.)

Define the *Lagrangian L* to be

$$L = \text{Kinetic energy} - \text{Potential energy}$$

$$= T - U.$$

We need to spend a bit of time on understanding the Lagrangian. It is here that the notation can become not only cumbersome but actually confusing, as variables will sometimes be independent variables and sometimes dependent variables, depending on context.

The kinetic energy is a function of the velocity while the potential energy is a function of the position (and for now the time). We can write the Lagrangian as

$$L \left(t, x, y, z, \frac{dx}{dt}, \frac{dy}{dt}, \frac{dz}{dt} \right) = T \left(\frac{dx}{dt}, \frac{dy}{dt}, \frac{dz}{dt} \right) - U(t, x, y, z).$$

Frequently people write derivatives with respect to time by putting a dot over the variable. With this notation,

$$\dot{x} = \frac{dx}{dt}, \quad \dot{y} = \frac{dy}{dt}, \quad \dot{z} = \frac{dz}{dt}.$$

This notation goes back to Newton; it is clearly preferred by typesetters in the following computations. With this notation, the Lagrangian is

$$L(t, x, y, z, \dot{x}, \dot{y}, \dot{z}).$$

Now to link this Lagrangian to mechanics. Suppose we are given two points in space:

$$p_0 = (x_0, y_0, z_0)$$
$$p_1 = (x_1, y_1, z_1).$$

Our goal is to find the actual path

$$r(t) = (x(t), y(t), z(t))$$

that a particle will follow from a time t_0 to a later time t_1, subject to the initial and final conditions

$$r(t_0) = p_0, r(t_1) = p_1.$$

Given any possible path $r(t)$, the Lagrangian L is a function on this path, as each point of the path gives us positions and velocities. We still need one more definition. Given any possible path $r(t)$ going from point p_0 to p_1, let the *action S* be defined as

$$S(r(t)) = \int_{t_0}^{t_1} L \, dt$$
$$= \int_{t_0}^{t_1} L(t, x, y, z, \dot{x}, \dot{y}, \dot{z}) \, dt.$$

We can finally state our integral version of Newton's second law.

Newton's Second Law (Calculus of Variations Approach): A body following a path $r(t)$ from an initial point p_0 to a final point p_1 is a critical point for the action $S(r(t))$.

Of course we now want to link this statement to the differential equation approach to Newton's second law. We must link the path that is a critical value for

$$S(r(t)) = \int_{t_0}^{t_1} L \, dt$$

to the differential equation

$$F = m \cdot a(t).$$

Assume we have found our critical path $r(t)$. Then the Lagrangian must satisfy the Euler-Lagrange equations:

$$\frac{\partial L}{\partial x} - \frac{d}{dt}\frac{\partial L}{\partial \dot{x}} = 0$$

$$\frac{\partial L}{\partial y} - \frac{d}{dt}\frac{\partial L}{\partial \dot{y}} = 0$$

$$\frac{\partial L}{\partial z} - \frac{d}{dt}\frac{\partial L}{\partial \dot{z}} = 0.$$

In taking the partial derivatives, we are treating the Lagrangian L as a function of the seven variables $t, x, y, z, \dot{x}, \dot{y}, \dot{z}$, but, in taking the derivative $\frac{d}{dt}$, we are treating L as the function $L(t, x(t), y(t), z(t), \dot{x}(t), \dot{y}(t), \dot{z}(t))$ of the variable t alone.

Since

$$L = \text{Kinetic energy} - \text{Potential enery}$$

$$= \frac{1}{2}m(\dot{x}^2 + \dot{y}^2 + \dot{z}^2) - U(x, y, z),$$

the Euler-Lagrange equations become

$$-\frac{\partial U}{\partial x} - \frac{d(m\dot{x})}{dt} = 0$$

$$-\frac{\partial U}{\partial y} - \frac{d(m\dot{y})}{dt} = 0$$

$$-\frac{\partial U}{\partial z} - \frac{d(m\dot{z})}{dt} = 0.$$

Since the $\dot{x}, \dot{y}, \dot{z}$ are just another notation for a derivative, we have

$$-\frac{\partial U}{\partial x} = m\frac{d^2 x}{dt^2}$$

$$-\frac{\partial U}{\partial y} = m\frac{d^2 y}{dt^2}$$

$$-\frac{\partial U}{\partial z} = m\frac{d^2 z}{dt^2}.$$

Note that the right-hand side of the preceding is just the mass (a number) times the individual components of the acceleration vector,

$$m \cdot a = m \left(\frac{d^2x}{dt^2}, \frac{d^2y}{dt^2}, \frac{d^2z}{dt^2} \right),$$

making it start to look like the desired Force = mass times acceleration. The final step is to recall that the very definition of the potential energy function U is that it is a function whose gradient gives us the force. Thus the force is the vector field:

$$F = -\text{gradient}(U) = -\nabla U = \left(-\frac{\partial U}{\partial x}, -\frac{\partial U}{\partial y}, -\frac{\partial U}{\partial z} \right).$$

The Euler-Lagrange equations are just $F = ma$ in disguise.

6.4. Conservation of Energy from Lagrangians

In the popular world there is the idea of conservation of energy, usually expressed in the form that energy can be neither created nor destroyed. I have even heard this phrase used by someone trying to make the existence of spirits and ghosts sound scientifically plausible. While such attempts are silly, it is indeed the case that, in a well-defined sense, in classical Newtonian physics, energy is indeed conserved. Historically, finding the right definitions and concepts was somewhat tortuous [11]. Technically, in classical physics, it came to be believed that in a closed system, energy must be a constant. Richard Feynman, in section 4.1 of *The Feynman Lectures on Physics*, Volume I [21], gave an excellent intuitive description, comparing energy to children's blocks.

In the context of the last section, we will show that

$$\text{Kinetic Energy} + \text{Potential Energy}$$

is a constant, meaning that we want to show that

$$\frac{1}{2} m \left(\dot{x}^2 + \dot{y}^2 + \dot{z}^2 \right) + U(x, y, z)$$

is a constant. We will see, in the exercises, that this is a consequence of

Theorem 6.4.1. *Suppose that* $L = L(t, x, y, z, \dot{x}, \dot{y}, \dot{z})$ *is independent of t, meaning that*

$$\frac{\partial L}{\partial t} = 0.$$

When restricting to a critical path $(x(t), y(t), z(t))$, the function

$$\frac{\partial L}{\partial \dot{x}}\dot{x} + \frac{\partial L}{\partial \dot{y}}\dot{y} + \frac{\partial L}{\partial \dot{z}}\dot{z} - L$$

is a constant function.

Proof. To show that $\frac{\partial L}{\partial \dot{x}}\dot{x} + \frac{\partial L}{\partial \dot{y}}\dot{y} + \frac{\partial L}{\partial \dot{z}}\dot{z} - L$ is a constant, we will prove that

$$\frac{d}{dt}\left(\frac{\partial L}{\partial \dot{x}}\dot{x} + \frac{\partial L}{\partial \dot{y}}\dot{y} + \frac{\partial L}{\partial \dot{z}}\dot{z} - L\right) = 0.$$

This will be an exercise in the multi-variable chain rule, critically using the Euler-Lagrange equations.

We have

$$\frac{dL}{dt} = \frac{\partial L}{\partial t}\frac{dt}{dt}$$

$$+ \frac{\partial L}{\partial x}\frac{dx}{dt} + \frac{\partial L}{\partial y}\frac{dy}{dt} + \frac{\partial L}{\partial z}\frac{dz}{dt}$$

$$+ \frac{\partial L}{\partial \dot{x}}\frac{d\dot{x}}{dt} + \frac{\partial L}{\partial \dot{y}}\frac{d\dot{y}}{dt} + \frac{\partial L}{\partial \dot{z}}\frac{d\dot{z}}{dt}.$$

Using our assumption that $\partial L/\partial t = 0$ and the Euler-Lagrange equations, we get

$$\frac{dL}{dt} = \frac{d}{dt}\left(\frac{\partial L}{\partial \dot{x}}\right)\dot{x} + \frac{d}{dt}\left(\frac{\partial L}{\partial \dot{y}}\right)\dot{y} + \frac{d}{dt}\left(\frac{\partial L}{\partial \dot{z}}\right)\dot{z}$$

$$+ \frac{\partial L}{\partial \dot{x}}\frac{d\dot{x}}{dt} + \frac{\partial L}{\partial \dot{y}}\frac{d\dot{y}}{dt} + \frac{\partial L}{\partial \dot{z}}\frac{d\dot{z}}{dt}$$

$$= \frac{d}{dt}\left(\frac{\partial L}{\partial \dot{x}}\dot{x} + \frac{\partial L}{\partial \dot{y}}\dot{y} + \frac{\partial L}{\partial \dot{z}}\dot{z}\right),$$

which gives us our desired $\frac{d}{dt}\left(\frac{\partial L}{\partial \dot{x}}\dot{x} + \frac{\partial L}{\partial \dot{y}}\dot{y} + \frac{\partial L}{\partial \dot{z}}\dot{z} - L\right) = 0.$

□

To finish this section, we need to link this with energy, as is done in the following corollary, whose proof is left to the exercises:

Corollary 6.4.1. *If*

$$L = \frac{m}{2}(\dot{x}^2 + \dot{y}^2 + \dot{z}^2) - U(x, y, z),$$

then the energy

$$\frac{m}{2}(\dot{x}^2 + \dot{y}^2 + \dot{z}^2) + U(x, y, z)$$

is a constant.

6.5. Noether's Theorem and Conservation Laws

Noether's Theorem is one of the most important results of the twentieth century. It provides a link among mechanics (as described by Lagrangians), symmetries (and hence group theory), and conservation laws. In essence, Noether's Theorem states that any time there is a continuous change of coordinates that leaves the Lagrangian alone, then there must be a quantity that does not change (which, in math language, means that there is an invariant, while in physics language, it means there is a conservation law). For example, in the last section, under the assumption that the Lagrangian does not change under a change of the time coordinate (which is basically what our assumption that $\partial L/\partial t = 0$ means), we derived the conservation of energy. This is an extremely special case of Noether's Theorem.

A general, sweeping statement of Noether's Theorem, which we quote from section 1.1 of Zeidler's *Quantum Field Theory I: Basics in Mathematics and Physics* [70] is

Conservation laws in physics are caused by symmetries of physical systems.

Hence, in classical mechanics, the conservation of energy is derivable from the symmetry of time change. Similarly, in classical mechanics, whenever the Lagrangian does not change under spatial translations, the corresponding quantity that does not change can be shown to be the momentum, and, whenever the Lagrangian does not change under rotations of the spatial coordinates, the corresponding conserved quantity can be shown to be the angular momentum. In special relativity, time and space are not independent of each other, so it should not be surprising, under the influence of Noether, that energy and momentum might not be conserved anymore. But, instead of energy and momentum being conserved, there is a four-vector (called the *energy-momentum four-vector*) that is conserved.

Here is the big picture. Different theories of mechanics are described by different Lagrangians. Different Lagrangians are invariant under different changes of coordinates (i.e., different symmetries). Different symmetries give rise to different conservation laws.

Which is more important; which is more basic: the Lagrangian, the symmetries, or the conservation laws? The answer is that all three are equally important, and to a large extent, imply each other. We could have built this book around any of the three.

There are two recent books on Noether's Theorem: Kosmann-Schwarzbach's *The Noether Theorems: Invariants and Conservation Laws in the Twentieth Century* [35] and Neuenschwander's *Emmy Noether's Wonderful Theorem* [47].

6.6. Exercises

Exercise 6.6.1. *(This is problem 1 in section 48 of [59].) Find the Euler-Lagrange equations for*

$$\int \frac{\sqrt{1+(x')^2}}{x}\,dt.$$

Exercise 6.6.2. *(This is problem 2 in section 48 of [59].) Find the extremal solutions to*

$$\int_0^4 \left(tx' - (x')^2 \right) dt.$$

Exercise 6.6.3. *Show that an extremum $(x(t), y(t))$ to*

$$\int_{t_0}^{t_1} f(t,x(t),x'(t),y(t),y'(t))dt,$$

subject to the initial conditions

$$x(t_0) = x_0, y(t_0) = y_0, x(t_1) = x_1, y(t_1) = y_1,$$

must satisfy

$$\frac{\partial}{\partial x} f(t,x(t),x'(t),y(t),y'(t)) - \frac{d}{dt}\frac{\partial}{\partial x'} f(t,x(t),x'(t),y(t),y'(t)) = 0$$

and

$$\frac{\partial}{\partial y} f(t,x(t),x'(t),y(t),y'(t)) - \frac{d}{dt}\frac{\partial}{\partial y'} f(t,x(t),x'(t),y(t),y'(t)) = 0.$$

Exercise 6.6.4. *Fix two points (t_0, x_0, y_0) and (t_1, x_1, y_1). Given any function $(x(t), y(t))$ with*

$$x(t_0) = x_0 \text{ and } x(t_1) = x_1$$

$$y(t_0) = y_0 \text{ and } y(t_1) = y_1,$$

we know that the arclength is given by the integral

$$\int_{t_0}^{t_1} \sqrt{1+(x')^2+(y')^2}dt.$$

Show that the shortest distance between these two points is indeed a straight line.

Exercise 6.6.5. *Let R be a rectangle in the (s,t)-plane, with boundary $\partial(R)$. Show that if the function $(x(s,t))$ minimizes the integral*

$$\int_R f\left(s,t,x(s,t),\frac{\partial x}{\partial s},\frac{\partial x}{\partial t} \right) ds dt,$$

then $x(s,t)$ will satisfy

$$\frac{\partial f}{\partial x} - \frac{\partial}{\partial s}\frac{\partial f}{\partial x_s} - \frac{\partial}{\partial t}\frac{\partial f}{\partial x_t} = 0.$$

(Here we are using the notation

$$x_s = \frac{\partial x}{\partial s}$$

$$x_t = \frac{\partial x}{\partial t}.)$$

Exercise 6.6.6. *Let R be a rectangle in the (s,t)-plane, with boundary $\partial(R)$. Show that if the functions $(x(s,t), y(s,t))$ minimize the integral*

$$\int_R f\left(s,t,x(s,t),\frac{\partial x}{\partial s},\frac{\partial x}{\partial t},y(s,t),\frac{\partial y}{\partial s},\frac{\partial y}{\partial t}\right)dsdt,$$

then $(x(s,t), y(s,t))$ will satisfy

$$\frac{\partial f}{\partial x} - \frac{\partial}{\partial s}\frac{\partial f}{\partial x_s} - \frac{\partial}{\partial t}\frac{\partial f}{\partial x_t} = 0$$

and

$$\frac{\partial f}{\partial y} - \frac{\partial}{\partial s}\frac{\partial f}{\partial y_s} - \frac{\partial}{\partial t}\frac{\partial f}{\partial y_t} = 0.$$

(Here we are using the notation

$$x_s = \frac{\partial x}{\partial s},\ y_s = \frac{\partial y}{\partial s}$$

$$x_t = \frac{\partial x}{\partial t},\ y_t = \frac{\partial y}{\partial t}.)$$

Exercise 6.6.7. *If the Lagrangian is*

$$L = \frac{m}{2}(\dot{x}^2 + \dot{y}^2 + \dot{z}^2) - U(x,y,z),$$

show that the energy

$$\frac{m}{2}(\dot{x}^2 + \dot{y}^2 + \dot{z}^2) + U(x,y,z)$$

is a constant.

Exercise 6.6.8. *Suppose that we have a particle whose mass $m = 2$, for some choice of units. Suppose that the potential energy is $U(x) = x^2$. For the Lagrangian $L = \dot{x}^2 - x^2$, explicitly solve the Euler-Lagrange equation. Use this solution to show explicitly that the energy $\dot{x}^2 + x^2$ is a constant. (Hint: In finding solutions for the Euler-Lagrange equation, think trig functions.)*

7

Potentials

Summary: The different roles of the scalar potential function and the vector potential function will be developed. Initially both potentials seem to be secondary mathematical artifacts, as compared to the seemingly more fundamental electric field $E(x,y,z,t)$ and magnetic field $B(x,y,z,t)$. This is not the case. Both potentials are critical for the Lagrangian description of Maxwell's equations (which is the goal of the next chapter) and, ultimately more importantly, are critical for seeing how to generalize Maxwell's equations to the strong and weak forces.

7.1. Using Potentials to Create Solutions for Maxwell's Equations

It is not at all clear how easy it is to find vector fields E, B, and j and a function ρ that satisfy the many requirements needed for Maxwell's equations. There is a mathematical technique that almost seems to be a trick that will allow us easily to construct such solutions; we will see, though, that this approach is far more than a mere mathematical artifice. We start with the trick.

Let $\phi(x,y,z,t)$ be any function and let $A(x,y,z,t)$ be any vector field $(A_1(x,y,z,t), A_2(x,y,z,t), A_3(x,y,z,t))$. We now simply set

$$E(x,y,z,t) = -\nabla\phi - \frac{\partial A(x,y,z,t)}{\partial t}$$

$$= -\left(\frac{\partial\phi}{\partial x}, \frac{\partial\phi}{\partial y}, \frac{\partial\phi}{\partial z}\right) - \left(\frac{\partial A_1}{\partial t}, \frac{\partial A_2}{\partial t}, \frac{\partial A_3}{\partial t}\right)$$

$$B(x,y,z,t) = \nabla \times A(x,y,z,t)$$

$$= \left(\frac{\partial A_3}{\partial y} - \frac{\partial A_2}{\partial z}, -\left(\frac{\partial A_3}{\partial x} - \frac{\partial A_1}{\partial z}\right), \frac{\partial A_2}{\partial x} - \frac{\partial A_1}{\partial y}\right).$$

By also setting
$$\rho = \mathrm{div}(E)$$
and
$$j = c^2\mathrm{curl}(B) - \frac{\partial E}{\partial t},$$
we get solutions to Maxwell's equations. (This is explicitly shown in the exercises.)

We call ϕ the *scalar potential* and A the *vector potential*. These potentials generate a lot of solutions. Of course, there is no reason to think if we have a real electric field E and a real magnetic field B then there should be these auxiliary potentials. The goal of the next section is to show that they actually exist.

(First for a few technical caveats. Throughout this book, we are assuming that all functions can be differentiated as many times as needed. We also assume that our functions are such that we can interchange derivatives and integrals when needed. In this section, we are critically using that the domains for our functions are $\mathbb{R}^3$, $\mathbb{R}^4$, or, at worst, contractible spaces, which is a concept from topology. If the domains are not contractible, then the results of this chapter are no longer true. In particular, there is no guarantee for the existence of potentials. The full discussion of these issues requires a more detailed knowledge of differential topology, such as in the classic *Differential Topology* by Guillemin and Pollack [30].)

7.2. Existence of Potentials

Our goal is to prove

Theorem 7.2.1. *Let E and B be vector fields satisfying $\nabla \times E = -\frac{\partial B}{\partial t}$ and $\nabla \cdot B = 0$. Then there exist a scalar potential function ϕ and a vector potential field A such that*

$$E(x,y,z,t) = -\nabla\phi - \frac{\partial A(x,y,z,t)}{\partial t}$$
$$B(x,y,z,t) = \nabla \times A(x,y,z,t).$$

Note that this theorem applies only to vector fields E and B satisfying Maxwell's equations.

To some extent, the existence of these potentials stems from the following theorem from vector calculus:

Theorem 7.2.2. *Let F be a vector field.*

1. *If $\nabla \times F = 0$, then there exists a function ϕ, the scalar potential, such that*

$$F(x,y,z,t) = \nabla \phi(x,y,z,t).$$

2. *If $\nabla \cdot F = 0$, then there exists a vector field A, the vector potential, such that*

$$F(x,y,z,t) = \nabla \times A(x,y,z,t).$$

(We give intuitions and a proof of this theorem in the Appendix at the end of this chapter.) Though this theorem will provide a fairly quick proof that potentials exist, it makes their existence feel more like a mathematical trick. This is one of the reasons why for many years potentials were viewed as not as basic as the fields E and B. As we will see in later chapters, this view is not correct.

Now for the proof of Theorem 7.2.1.

Proof. Since we are assuming $\nabla \cdot B = 0$, we know from the preceding theorem that there is a vector field A such that

$$B = \nabla \times A.$$

We must now show that the function ϕ exists. We know that

$$\nabla \times E = -\frac{\partial B}{\partial t}$$

which means that

$$\nabla \times E = -\frac{\partial (\nabla \times A)}{\partial t}.$$

As will be shown in the exercises, this can be rewritten as

$$\nabla \times E = -\nabla \times \frac{\partial A}{\partial t},$$

meaning that

$$0 = \nabla \times E + \nabla \times \frac{\partial A}{\partial t}$$

$$= \nabla \times \left(E + \frac{\partial A}{\partial t} \right).$$

But then there must be a function ϕ such that

$$E + \frac{\partial A}{\partial t} = -\nabla \cdot \phi$$

giving us that

$$E = -\nabla \cdot \phi - \frac{\partial A}{\partial t}.$$

$\square$

7.3. Ambiguity in the Potential

In any given physical situation, there is no ambiguity in the electric field E and the magnetic field B. This is not the case with the potentials. The goal of this section is to start to understand the ambiguity of the potentials, namely,

Theorem 7.3.1. *Let E and B be vector fields satisfying Maxwell's equations, with scalar potential function ϕ and vector potential field A. If $f(x,y,z,t)$ is any function, then the function $\phi - \frac{\partial f}{\partial t}$ and the vector field $A + \nabla(f)$ are also potentials for E and B.*

The proof is a straightforward argument from vector calculus and is in the exercises.

Note that even using the phrase "The proof is a straightforward argument from vector calculus" makes this result seem not very important, suggesting even further that the potentials are just not as basic as the electric field E and the magnetic field B. Again, this is not the case. In later chapters we will see that potentials have a deep interpretation in modern differential geometry. (Technically, potentials are connections in a certain vector bundle.)

Also, in quantum mechanics, there is the Aharonov-Bohm effect, which proves that potentials have just as much status as the original fields E and B.

7.4. Appendix: Some Vector Calculus

We will now proceed to prove Theorem 7.2.2. Let

$$F = (F_1, F_2, F_3)$$

be a vector field. We first assume that its curl is zero: $\nabla \times F = (0,0,0)$, which means that

$$\frac{\partial F_1}{\partial y} = \frac{\partial F_2}{\partial x}$$

$$\frac{\partial F_1}{\partial z} = \frac{\partial F_3}{\partial x}$$

$$\frac{\partial F_3}{\partial y} = \frac{\partial F_2}{\partial z}.$$

We want to find a function ϕ such that $\nabla\phi = F$, meaning we want

$$\frac{\partial\phi}{\partial x} = F_1$$

$$\frac{\partial\phi}{\partial y} = F_2$$

$$\frac{\partial\phi}{\partial z} = F_3.$$

We will just state how to construct the function ϕ, ignoring, unfortunately, any clue as to how anyone could initially think up this function. Given a point $(x,y,z) \in \mathbb{R}^3$, consider the straight line γ from the origin $(0,0,0)$ to (x,y,z) given by

$$r(t) = (xt, yt, zt),$$

for $0 \le t \le 1$. as in Figure 7.1. Then define

$$\phi(x,y,z) = \int_\gamma F \cdot dr.$$

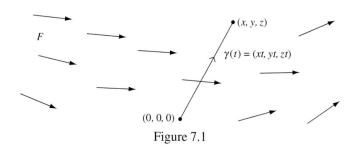

Figure 7.1

We have

$$\phi(x,y,z) = \int_\gamma F \cdot dr$$

$$= \int_0^1 (x F_1(tx,ty,tz) + y F_2(tx,ty,tz) + z F_3(tx,ty,tz))\, dt.$$

We will now show that $\partial\phi/\partial x = F_1$, leaving $\partial\phi/\partial y = F_2$ for the exercises (showing $\partial\phi/\partial z = F_3$ is similar). We have

$$\frac{\partial\phi}{\partial x} = \frac{\partial}{\partial x}\left(\int_0^1 (xF_1(tx,ty,tz) + yF_2(tx,ty,tz) + zF_3(tx,ty,tz))\,dt\right)$$

$$= \int_0^1 \frac{\partial}{\partial x}(xF_1(tx,ty,tz) + yF_2(tx,ty,tz) + zF_3(tx,ty,tz))\,dt$$

$$= \int_0^1 \left(F_1(tx,ty,tz) + tx\frac{\partial F_1}{\partial x} + ty\frac{\partial F_2}{\partial x} + tz\frac{\partial F_3}{\partial x}\right)dt$$

$$= \int_0^1 \left(F_1(tx,ty,tz) + tx\frac{\partial F_1}{\partial x} + ty\frac{\partial F_1}{\partial y} + tz\frac{\partial F_1}{\partial z}\right)dt$$

(using $\nabla \times F = 0$)

$$= \int_0^1 \left(F_1(tx,ty,tz) + t\frac{d}{dt}(F_1(tx,ty,tz))\right)dt.$$

Using integration by parts, we have

$$\int_0^1 t\frac{d}{dt}(F_1(tx,ty,tz))\,dt = F_1(x,y,z) - \int_0^1 F_1(tx,ty,tz)\,dt.$$

Then

$$\frac{\partial\phi}{\partial x} = \int_0^1 \left(F_1(tx,ty,tz) + t\frac{d}{dt}(F_1(tx,ty,tz))\right)dt = F_1(x,y,z),$$

as desired.

Now for the second part of the theorem. Assume that $\nabla \cdot F = 0$, meaning that

$$\frac{\partial F_1}{\partial x} + \frac{\partial F_2}{\partial y} + \frac{\partial F_3}{\partial z} = 0.$$

(Eventually we will use that $\frac{\partial F_1}{\partial x} = -\frac{\partial F_2}{\partial y} - \frac{\partial F_3}{\partial z}$.) We must construct a vector field $G = (G_1, G_2, G_3)$ such that $\nabla \times G = F$, or, in other words, such that

$$F_1 = \frac{\partial G_3}{\partial y} - \frac{\partial G_2}{\partial z}$$

$$F_2 = \frac{\partial G_1}{\partial z} - \frac{\partial G_3}{\partial x}$$

$$F_3 = \frac{\partial G_2}{\partial x} - \frac{\partial G_1}{\partial y}.$$

Set

$$G_1(x,y,z) = \int_0^1 (zt\,F_2(tx,ty,tz) - yt\,F_3(tx,ty,tz))\,dt$$

$$G_2(x,y,z) = \int_0^1 (xt\,F_3(tx,ty,tz) - zt\,F_1(tx,ty,tz))\,dt$$

$$G_3(x,y,z) = \int_0^1 (yt\,F_1(tx,ty,tz) - xt\,F_2(tx,ty,tz))\,dt.$$

(Again, we are giving no motivation for how anyone first thought of this method.) We will show that

$$F_1 = \frac{\partial G_3}{\partial y} - \frac{\partial G_2}{\partial z},$$

leaving the other parts for the exercises. We have

$$\frac{\partial G_3}{\partial y} - \frac{\partial G_2}{\partial z} = \frac{\partial}{\partial y}\left(\int_0^1 (yt\,F_1(tx,ty,tz) - xt\,F_2(tx,ty,tz))\,dt\right)$$

$$-\frac{\partial}{\partial z}\left(\int_0^1 (xt\,F_3(tx,ty,tz) - zt\,F_1(tx,ty,tz))\,dt\right)$$

$$= \int_0^1 \frac{\partial}{\partial y}(yt\,F_1(tx,ty,tz) - xt\,F_2(tx,ty,tz))\,dt$$

$$-\int_0^1 \frac{\partial}{\partial z}(xt\,F_3(tx,ty,tz) - zt\,F_1(tx,ty,tz))\,\mathrm{d}t$$

$$=\int_0^1 \left(t\,F_1 + yt^2\frac{\partial F_1}{\partial y} - xt^2\frac{\partial F_2}{\partial y}\right)\mathrm{d}t$$

$$-\int_0^1 \left(xt^2\frac{\partial F_3}{\partial z} - t\,F_1 - zt^2\frac{\partial F_1}{\partial z}\right)\mathrm{d}t.$$

Since $\frac{\partial F_1}{\partial x} = -\frac{\partial F_2}{\partial y} - \frac{\partial F_3}{\partial z}$, we have that

$$-xt^2\frac{\partial F_2}{\partial y} - xt^2\frac{\partial F_3}{\partial z} = xt^2\frac{\partial F_1}{\partial x}.$$

Then the preceding becomes

$$\frac{\partial G_3}{\partial y} - \frac{\partial G_2}{\partial z} = \int_0^1 \left(2t\,F_1 + t^2\frac{\mathrm{d}}{\mathrm{d}t}F_1(tx,ty,tz)\right)\mathrm{d}t.$$

Using integration by parts we have

$$\int_0^1 t^2\frac{\mathrm{d}}{\mathrm{d}t}F_1(tx,ty,tz)\,\mathrm{d}t = t^2 F_1(tx,ty,tz)\Big|_0^1 - \int_0^1 2t\,F_1\,\mathrm{d}t = F_1 - \int_0^1 2t\,F_1\,\mathrm{d}t.$$

Thus

$$\frac{\partial G_3}{\partial y} - \frac{\partial G_2}{\partial z} = F_1(x,y,z),$$

as desired.

7.5. Exercises

Exercise 7.5.1. *Let*

$$\phi(x,y,z,t) = xyzt$$

and

$$A(x,y,z,t) = (x^2 y + t^3, xyz^2 t, t^3 x + yzt).$$

Compute

$$E(x,y,z,t) = -\nabla\phi - \frac{\partial A(x,y,z,t)}{\partial t}$$

$$B(x,y,z,t) = \nabla \times A(x,y,z,t).$$

Then compute

$$\rho = \text{div}(E)$$

$$j = c^2 \text{curl}(B) - \frac{\partial E}{\partial t}.$$

Verify that these are solutions to Maxwell's equations.

Exercise 7.5.2. *Let $\phi(x,y,z,t)$ be any function and $A(x,y,z,t)$ any vector field. Setting*

$$E(x,y,z,t) = -\nabla\phi - \frac{\partial A(x,y,z,t)}{\partial t}$$

$$B(x,y,z,t) = \nabla \times A(x,y,z,t)$$

and then in turn setting

$$\rho = \text{div}(E)$$

$$j = c^2 \text{curl}(B) - \frac{\partial E}{\partial t},$$

show that we have a solution to Maxwell's equations.

Exercise 7.5.3. *Let*

$$f(x,y,z,t) = x^2 + yz + zt^2.$$

Using the notation in the first exercise of this chapter, replace the preceding ϕ with

$$\phi - \frac{\partial f}{\partial t}$$

and A with

$$A + \nabla(f).$$

Show that we get the same values as in the first exercise for $E, B, j,$ and ρ.

Exercise 7.5.4. *For any vector field $F = (F_1, F_2, F_3)$, show that*

$$\frac{\partial(\nabla \times F)}{\partial t} = \nabla \times \frac{\partial F}{\partial t}.$$

Exercise 7.5.5. *For $\mathbf{F} = (F_1, F_2, F_3)$, show that*

$$\nabla \times \mathbf{F} = 0$$

if and only if

$$\frac{\partial F_2}{\partial x} = \frac{\partial F_1}{\partial y}$$

$$\frac{\partial F_3}{\partial x} = \frac{\partial F_1}{\partial z}$$

$$\frac{\partial F_2}{\partial z} = \frac{\partial F_3}{\partial y}.$$

Exercise 7.5.6. *Prove that if E and B are any vector fields satisfying Maxwell's equations, with potential function ϕ and potential vector field A, then, for any function $f(x,y,z,t)$, the function $\phi - \frac{\partial f}{\partial t}$ and the vector field $A + \nabla(f)$ are also potentials for E and B.*

Exercise 7.5.7. *Let F be a vector field with $\nabla \times F = (0,0,0)$. Set*

$$\phi(x,y,z) = \int_0^1 (x F_1(tx,ty,tz) + y F_2(tx,ty,tz) + z F_3(tx,ty,tz))\, dt.$$

Show that

$$\frac{\partial \phi}{\partial y} = F_2.$$

Exercise 7.5.8. *Let*

$$F = (x,y,z).$$

Show that $\nabla \times F = (0,0,0)$ and then find a function $\phi(x,y,z)$ such that

$$\nabla \cdot \phi = F.$$

Exercise 7.5.9. *Let F be a vector field such that $\nabla \cdot F = 0$. Set*

$$G_1(x,y,z) = \int_0^1 (zt F_2(tx,ty,tz) - yt F_3(tx,ty,tz))\, dt$$

$$G_3(x,y,z) = \int_0^1 (yt F_1(tx,ty,tz) - xt F_2(tx,ty,tz))\, dt.$$

Show that

$$F_2 = \frac{\partial G_1}{\partial z} - \frac{\partial G_3}{\partial x}.$$

Exercise 7.5.10. *Let $F = (yz,x,y)$. Show that $\nabla \cdot F = 0$. Then find a vector field G such that*

$$\nabla \times G = F.$$

8

Lagrangians and Electromagnetic Forces

Summary: Using last chapter's scalar and vector potentials, we will construct the Lagrangian whose corresponding Euler-Lagrange equation

$$F = q(E + v \times B)$$

gives us the electromagnetic force F.

8.1. Desired Properties for the Electromagnetic Lagrangian

In Chapter 6, we saw how to recast basic mechanics in terms of a Lagrangian. We want to see how to do this for electromagnetic forces. Thus we want to find a function

$$L = L(t, x, y, z, \dot{x}, \dot{y}, \dot{z})$$

with the following key property. Suppose we have a particle moving in an electromagnetic field that starts at the point (x_0, y_0, z_0) and ends at the point (x_1, y_1, z_1). We want to be able to predict the path $(x(t), y(t), z(t))$ for $0 \leq t \leq 1$ that the particle follows. Thus we want our function L to have the property that, among all possible paths, the actual path is a critical point of the integral

$$\int_0^1 L(t, x, y, z, \dot{x}, \dot{y}, \dot{z}) \, dt.$$

From Chapter 5, we know that the particle's path $(x(t), y(t), z(t))$ will satisfy the system of differential equations

$$F = q(E + v \times B).$$

98

From Chapter 6, we know for this path that the Euler-Lagrange equations

$$\frac{d}{dt}\left(\frac{\partial L}{\partial \dot{x}}\right) - \frac{\partial L}{\partial x} = 0$$

$$\frac{d}{dt}\left(\frac{\partial L}{\partial \dot{y}}\right) - \frac{\partial L}{\partial y} = 0$$

$$\frac{d}{dt}\left(\frac{\partial L}{\partial \dot{z}}\right) - \frac{\partial L}{\partial z} = 0$$

must be satisfied. A candidate function $L(t,x,y,z,\dot{x},\dot{y},\dot{z})$ will be our desired Lagrangian if the Euler-Lagrange equations can be shown to be equivalent to $F = q(E + v \times B)$.

One final word about notation, which can quickly become a nightmare. When we write $L = L(t,x,y,z,\dot{x},\dot{y},\dot{z})$, we are thinking of L as a function of seven independent variables. The reason to denote the last three variables as $\dot{x}, \dot{y}, \dot{z}$ instead of, say, u, v, w, is that when we find our path $(x(t), y(t), z(t))$ that minimizes $\int_0^1 L(t,x,y,z,\dot{x},\dot{y},\dot{z})dt$, we will have

$$\dot{x} = \frac{dx(t)}{dt}$$

$$\dot{y} = \frac{dy(t)}{dt}$$

$$\dot{z} = \frac{dz(t)}{dt}.$$

8.2. The Electromagnetic Lagrangian

The Lagrangian for electromagnetism is

$$L = \frac{1}{2}mv \cdot v - q\phi + qA \cdot v.$$

The vector v denotes the velocity $v = (\dot{x}, \dot{y}, \dot{z})$, the function $\phi(x,y,z,t)$ is the scalar potential, the vector field $A(x,y,z,t)$ is the vector potential. Also, m is the mass of the particle and q is its charge, both of which are constants. Thus this L is a function of the variables $t,x,y,z,\dot{x},\dot{y},\dot{z}$.

We are simply stating the Lagrangian L and giving no clue as to how anyone could have ever first written it down. Our justification will be that the Euler-Lagrange equations for this L will yield the correct force $F = q(E + v \times B)$ for electromagnetism. Hence the following:

Theorem 8.2.1. *Any path $(x(t), y(t), z(t))$, for $0 \le t \le 1$, that minimizes*

$$\int_0^1 L(t, x, y, z, \dot{x}, \dot{y}, \dot{z}) \, dt$$

will satisfy the partial differential equations given by

$$F = q(E + v \times B).$$

Proof. The ordinary differential equations given by $F = q(E + v \times B)$ are

$$m\ddot{x} = q E_1 + q(\dot{y} B_3 - \dot{z} B_2)$$
$$m\ddot{y} = q E_2 + q(\dot{z} B_1 - \dot{x} B_3)$$
$$m\ddot{z} = q E_3 + q(\dot{x} B_2 - \dot{y} B_1).$$

Similar to $\dot{x}$ denoting the derivative of $x(t)$ with respect to time t, we are letting $\ddot{x}$ denote the second derivative of $x(t)$ with respect to time t, and similar for $\ddot{y}$, $\ddot{z}$, and so on.

We must show that these are implied by the Euler-Lagrange equations:

$$\frac{d}{dt}\left(\frac{\partial L}{\partial \dot{x}}\right) - \frac{\partial L}{\partial x} = 0$$

$$\frac{d}{dt}\left(\frac{\partial L}{\partial \dot{y}}\right) - \frac{\partial L}{\partial y} = 0$$

$$\frac{d}{dt}\left(\frac{\partial L}{\partial \dot{z}}\right) - \frac{\partial L}{\partial z} = 0.$$

We will show here that $\frac{d}{dt}\left(\frac{\partial L}{\partial \dot{x}}\right) - \frac{\partial L}{\partial x} = 0$ implies that $m\ddot{x} = q E_1 + q(\dot{y} B_3 - \dot{z} B_2)$, leaving the rest to the exercises.

We are assuming that the Lagrangian is

$$L = \frac{1}{2} m v \cdot v - q\phi + q A \cdot v.$$

The scalar potential ϕ and the vector potential $A = (A_1, A_2, A_3)$ are functions of x, y, z and do not directly depend on the derivatives $\dot{x}, \dot{y}, \dot{z}$. The velocity v, of course, is just $v = (\dot{x}, \dot{y}, \dot{z})$. Thus the Lagrangian is

$$L(t, x, y, z, \dot{x}, \dot{y}, \dot{z}) = \frac{1}{2} m(\dot{x}^2 + \dot{y}^2 + \dot{z}^2) - q\phi(x, y, z, t) + q(A_1 \dot{x} + A_2 \dot{y} + A_3 \dot{z}).$$

Then

$$\frac{\partial L}{\partial \dot{x}} = m\dot{x} + q A_1.$$

Hence,

$$\frac{d}{dt}\left(\frac{\partial L}{\partial \dot{x}}\right) = m\ddot{x} + q\,\frac{dA_1(t,x(t),y(t),z(t))}{dt}$$

$$= m\ddot{x} + q\left(\frac{\partial A_1}{\partial t}\cdot\frac{dt}{dt} + \frac{\partial A_1}{\partial x}\cdot\frac{dx}{dt} + \frac{\partial A_1}{\partial y}\cdot\frac{dy}{dt} + \frac{\partial A_1}{\partial z}\cdot\frac{dz}{dt}\right)$$

$$= m\ddot{x} + q\left(\frac{\partial A_1}{\partial t} + \frac{\partial A_1}{\partial x}\dot{x} + \frac{\partial A_1}{\partial y}\dot{y} + \frac{\partial A_1}{\partial z}\dot{z}\right).$$

Now,

$$\frac{\partial L}{\partial x} = -q\,\frac{\partial\phi}{\partial x} + q\left(\frac{\partial A_1}{\partial x}\dot{x} + \frac{\partial A_2}{\partial x}\dot{y} + \frac{\partial A_3}{\partial x}\dot{z}\right).$$

Then the Euler-Lagrange equation

$$\frac{\partial L}{\partial x} - \frac{d}{dt}\left(\frac{\partial L}{\partial \dot{x}}\right) = 0$$

becomes

$$m\ddot{x} = -q\,\frac{\partial\phi}{\partial x} - q\,\frac{\partial A_1}{\partial t} + q\left[\dot{y}\left(\frac{\partial A_2}{\partial x} - \frac{\partial A_1}{\partial y}\right) + \dot{z}\left(\frac{\partial A_3}{\partial x} - \frac{\partial A_1}{\partial z}\right)\right],$$

which is the desired formula

$$m\ddot{x} = q\,E_1 + q\,(\dot{y}\,B_3 - \dot{z}\,B_2).$$

□

There is still the question of how anyone ever came up with the Lagrangian $L = \frac{1}{2}mv\cdot v - q\phi + qA\cdot v$. In general, given a system of differential equations, one can always ask whether there is a Lagrangian whose minimal paths give solutions to the initial equations. This is called the Inverse Problem for Lagrangians. Of course, for any specific system, such as for our system $F = q(E + v\times B)$, the corresponding L can be arrived at via fiddling with the Euler-Lagrange equations, as is indeed what originally happened.

8.3. Exercises

Exercise 8.3.1. *Show that*

$$\frac{d}{dt}\left(\frac{\partial L}{\partial \dot{y}}\right) - \frac{\partial L}{\partial y} = 0$$

implies that

$$m\ddot{y} = q\,E_2 + \frac{q}{c}(\dot{z}\,B_1 - \dot{x}\,B_3)$$

and that

$$\frac{d}{dt}\left(\frac{\partial L}{\partial \dot{z}}\right) - \frac{\partial L}{\partial z} = 0$$

implies that

$$m\ddot{z} = q\,E_3 + q(\dot{x}\,B_2 - \dot{y}\,B_1).$$

Exercise 8.3.2. *Let $\phi(x,y,z) = xyz^2$ be a scalar potential and $A = (xyz, x + z, y + 3x)$ be a vector potential. Find the Lagrangian and then write down the corresponding Euler-Lagrange equations.*

Exercise 8.3.3. *Use the notation of the previous problem. Let $f(x,y,z) = x^2 + y^2z$. Find the Lagrangian for the scalar potential*

$$\phi - \frac{\partial f}{\partial t}$$

and vector potential

$$A + \nabla(f).$$

Compute the corresponding Euler-Lagrange equations, showing that these are the same as the Euler-Lagrange equations in the previous problem.

Exercise 8.3.4. *Let ϕ be a scalar potential, A be a vector potential, and f be any function. Show that the Euler-Lagrange equations are the same for the Lagrangian using the scalar potential ϕ and the vector potential A as for the Lagrangian using the scalar potential $\phi - \frac{\partial f}{\partial t}$ and vector potential $A + \nabla(f)$.*

9

Differential Forms

Summary: This chapter will develop the basic definitions of differential forms and the exterior algebra. The emphasis will be on how to compute with differential forms. The exterior algebra is both a highly efficient language for the basic terms from vector calculus and a language that can be easily generalized to far broader situations. Our eventual goal is to recast Maxwell's equations in terms of differential forms.

9.1. The Vector Spaces $\Lambda^k(\mathbb{R}^n)$

9.1.1. A First Pass at the Definition

(In this subsection we give a preliminary definition for k-forms. Here they will be finite-dimensional real vector spaces. In the next subsection we will give the official definition, where we extend the coefficients to allow for functions and not just numbers.)

Let $x_1, x_2, \ldots, x_n$ be coordinates for $\mathbb{R}^n$. We want to define the vector space of k-forms. For now, we will concentrate on explicit definitions and methods of manipulation. (Underlying intuitions will be developed in the next section.) This division reflects that knowing how to calculate with k-forms is to a large extent independent of what they mean. This is analogous to differentiation in beginning calculus, where we care about derivatives because of their meaning (i.e., we want to know a function's rate of change, or the slope of a tangent line) but use derivatives since they are easy to calculate (using various rules of differentiation, such as the product rule, chain rule, quotient rule, etc.). Thinking about the meaning of the derivative rarely aids in calculating.

For each nonnegative integer k, we will define a vector space of dimension

$$\binom{n}{k} = \frac{n!}{k!(n-k)!}.$$

We write down basis elements first. For each sequence of positive integers $1 \leq i_1 < i_2 < \cdots < i_k \leq n$, we write down the symbol

$$\mathrm{d}x_{i_1} \wedge \mathrm{d}x_{i_2} \wedge \cdots \wedge \mathrm{d}x_{i_k}$$

which we call an *elementary k-form*. For $I = \{i_1, i_2, \ldots i_k\}$, we write

$$\mathrm{d}x_I := \mathrm{d}x_{i_1} \wedge \mathrm{d}x_{i_2} \wedge \cdots \wedge \mathrm{d}x_{i_k}.$$

There are $\binom{n}{k}$ such symbols. Then we define, temporarily, $\Lambda^k(\mathbb{R}^n)$ to be the vector space obtained by taking finite linear combinations, using real numbers as coefficients, of these elementary k-forms. (In the next subsection we will extend the definition of $\Lambda^k(\mathbb{R}^n)$ by allowing functions to be coefficients.)

Let us consider the forms for $\mathbb{R}^3$, with coordinates x, y, z. The elementary 1-forms are

$$\mathrm{d}x, \mathrm{d}y, \mathrm{d}z,$$

the elementary 2-forms are

$$\mathrm{d}x \wedge \mathrm{d}y, \ \mathrm{d}x \wedge \mathrm{d}z, \ \mathrm{d}y \wedge \mathrm{d}z,$$

and the elementary 3-form is

$$\mathrm{d}x \wedge \mathrm{d}y \wedge \mathrm{d}z.$$

The elementary 0-form is denoted by 1.

Then the vector space of 1-forms is

$$\Lambda^1(\mathbb{R}^3) = \{a\mathrm{d}x + b\mathrm{d}y + c\mathrm{d}z : a, b, c \in \mathbb{R}\}.$$

This is a vector space if we define scalar multiplication by

$$\lambda(a\mathrm{d}x + b\mathrm{d}y + c\mathrm{d}z) = \lambda a\mathrm{d}x + \lambda b\mathrm{d}y + \lambda c\mathrm{d}z$$

and vector addition by

$$(a_1\mathrm{d}x + b_1\mathrm{d}y + c_1\mathrm{d}z) + (a_2\mathrm{d}x + b_2\mathrm{d}y + c_2\mathrm{d}z)$$
$$= (a_1 + a_2)\mathrm{d}x + (b_1 + b_2)\mathrm{d}y + (c_1 + c_2)\mathrm{d}z.$$

In a similar way, we have

$$\Lambda^0(\mathbb{R}^3) = \{a \cdot 1 : a \in \mathbb{R}\}$$
$$\Lambda^2(\mathbb{R}^3) = \{a\mathrm{d}x \wedge \mathrm{d}y + b\mathrm{d}x \wedge \mathrm{d}z + c\mathrm{d}y \wedge \mathrm{d}z : a, b, c \in \mathbb{R}\}$$
$$\Lambda^3(\mathbb{R}^3) = \{a\mathrm{d}x \wedge \mathrm{d}y \wedge \mathrm{d}z : a \in \mathbb{R}\}.$$

Often we write $\Lambda^0(\mathbb{R}^3)$ simply as $\mathbb{R}$.

The elementary k-form $\mathrm{d}x_{i_1} \wedge \mathrm{d}x_{i_2} \wedge \cdots \wedge \mathrm{d}x_{i_k}$ depends on the ordering $i_1 < i_2 < \cdots < i_k$. We want to make sense out of any

$$\mathrm{d}x_{i_1} \wedge \mathrm{d}x_{i_2} \wedge \cdots \wedge \mathrm{d}x_{i_k},$$

even if we do not have $i_1 < i_2 < \cdots < i_k$. The idea is that if we interchange any two of the terms in $\mathrm{d}x_{i_1} \wedge \mathrm{d}x_{i_2} \wedge \cdots \wedge \mathrm{d}x_{i_k}$, we change the sign. For example, we define

$$\mathrm{d}x_2 \wedge \mathrm{d}x_1 = -\mathrm{d}x_1 \wedge \mathrm{d}x_2$$

and

$$\mathrm{d}x_3 \wedge \mathrm{d}x_2 \wedge \mathrm{d}x_1 = -\mathrm{d}x_1 \wedge \mathrm{d}x_2 \wedge \mathrm{d}x_3.$$

In general, given a collection $(i_1, i_2, \ldots, i_k)$, there is a reordering σ (called a *permutation*) such that

$$\sigma(i_1) < \cdots < \sigma(i_k).$$

Each such permutation is the composition of interchangings of two terms. If we need an even number of such permuations we say that the *sign* of σ is 1, while if we need an odd number, the sign is -1. (The notion of "even/odd" is well-defined, since it can be shown that a given permutation cannot be both even and odd.) Then we define

$$\mathrm{d}x_{i_1} \wedge \mathrm{d}x_{i_2} \wedge \cdots \wedge \mathrm{d}x_{i_k} = \mathrm{sign}(\sigma) \cdot \mathrm{d}x_{\sigma(i_1)} \wedge \mathrm{d}x_{\sigma(i_2)} \wedge \cdots \wedge \mathrm{d}x_{\sigma(i_k)}.$$

Let us consider the 3-form $\mathrm{d}x_1 \wedge \mathrm{d}x_2 \wedge \mathrm{d}x_3$. There are $3! = 6$ ways for rearranging $(1,2,3)$, including the "rearrangement that leaves them alone." We have

$$\mathrm{d}x_1 \wedge \mathrm{d}x_2 \wedge \mathrm{d}x_3 = \mathrm{d}x_1 \wedge \mathrm{d}x_2 \wedge \mathrm{d}x_3$$

$$\mathrm{d}x_2 \wedge \mathrm{d}x_1 \wedge \mathrm{d}x_3 = -\mathrm{d}x_1 \wedge \mathrm{d}x_2 \wedge \mathrm{d}x_3$$

$$\mathrm{d}x_3 \wedge \mathrm{d}x_2 \wedge \mathrm{d}x_1 = -\mathrm{d}x_1 \wedge \mathrm{d}x_2 \wedge \mathrm{d}x_3$$

$$\mathrm{d}x_1 \wedge \mathrm{d}x_3 \wedge \mathrm{d}x_2 = -\mathrm{d}x_1 \wedge \mathrm{d}x_2 \wedge \mathrm{d}x_3$$

$$\mathrm{d}x_3 \wedge \mathrm{d}x_1 \wedge \mathrm{d}x_2 = -\mathrm{d}x_1 \wedge \mathrm{d}x_3 \wedge \mathrm{d}x_2$$

$$= \mathrm{d}x_1 \wedge \mathrm{d}x_2 \wedge \mathrm{d}x_3$$

$$\mathrm{d}x_2 \wedge \mathrm{d}x_3 \wedge \mathrm{d}x_1 = -\mathrm{d}x_2 \wedge \mathrm{d}x_1 \wedge \mathrm{d}x_3$$

$$= \mathrm{d}x_1 \wedge \mathrm{d}x_2 \wedge \mathrm{d}x_3.$$

Finally, we need to give meaning to symbols for $\mathrm{d}x_{i_1} \wedge \mathrm{d}x_{i_2} \wedge \cdots \wedge \mathrm{d}x_{i_k}$ when the $i_1, i_2, \ldots, i_k$ are not all distinct. We just declare this type of k-form to be zero. This intuitively agrees with what we just did, since if

$$\mathrm{d}x_i \wedge \mathrm{d}x_j = -\mathrm{d}x_j \wedge \mathrm{d}x_i,$$

then we should have, for $i = j$,

$$\mathrm{d}x_i \wedge \mathrm{d}x_i = -\mathrm{d}x_i \wedge \mathrm{d}x_i = 0.$$

9.1.2. Functions as Coefficients

We have defined each $\Lambda^k(\mathbb{R}^n)$ to be a real vector space. Thus the coefficients are real numbers. But we want to allow the coefficients to be functions. For example, we want to make sense out of terms such as

$$(x^2 + y)\mathrm{d}x + xyz\mathrm{d}y + yz^3\mathrm{d}z$$

and

$$\sin(xy^2z)\mathrm{d}x \wedge \mathrm{d}y + e^y\mathrm{d}x \wedge \mathrm{d}z + z\mathrm{d}y \wedge \mathrm{d}z.$$

Using tools from abstract algebra, the most sophisticated way is to start with a ring of functions on $\mathbb{R}^n$, such as the ring of differentiable functions. Then we now officially define each $\Lambda^k(\mathbb{R}^n)$ to be the module over the ring of functions with basis elements the elementary k-forms. This will be the definition that we will be using for $\Lambda^k(\mathbb{R}^n)$ for the rest of the book. (As is the case throughout this book, our functions are infinitely differentiable.)

There is a more down-to-earth approach, though. We want to make sense out of the symbol

$$\sum_I f_I(x_1, x_2, \ldots, x_n)\mathrm{d}x_I,$$

where we are summing over all indices $I = \{i_1, i_2, \ldots i_k\}$ for $i_1 < i_2 < \cdots < i_k$ and where the f_I are functions. For each point $a = (a_1, a_2, \ldots, a_n) \in \mathbb{R}^n$, we interpret $\sum_I f_I(x_1, x_2, \ldots, x_n)\mathrm{d}x_I$ at a as the k-form

$$\sum_I f_I(a)\mathrm{d}x_I = \sum_I f_I(a_1, a_2, \ldots, a_n)\mathrm{d}x_I.$$

Thus $\sum_I f_I(x_1, x_2, \ldots, x_n)\mathrm{d}x_I$ can be thought of as a way of defining a whole family of k-forms, changing at each point $a \in \mathbb{R}^n$.

9.1.3. The Exterior Derivative

So far, we have treated k-forms $\sum_I f_I(x_1, x_2, \ldots, x_n)\mathrm{d}x_I$ quite formally, with no interpretation yet given. Still, we have symbols like $\mathrm{d}x$ showing up. This suggests that underlying our eventual interpretation will be derivatives. For now, we simply start with the definition of the exterior derivative:

Definition 9.1.1. *The* exterior derivative

$$\mathrm{d} : \Lambda^k(\mathbb{R}^n) \to \Lambda^{k+1}(\mathbb{R}^n)$$

is defined by first setting, for any real-valued function $f(x_1, \ldots, x_n)$,

$$\mathrm{d}(f(x_1, x_2, \ldots, x_n)\mathrm{d}x_I) = \sum_{j=1}^{n} \frac{\partial f}{\partial x_j} \mathrm{d}x_j \wedge \mathrm{d}x_I$$

and then, for sums of elementary k-forms, setting

$$\mathrm{d}\left(\sum_I f_I(x_1, x_2, \ldots, x_n)\mathrm{d}x_I\right) = \sum_I \mathrm{d}(f_I(x_1, x_2, \ldots, x_n) \wedge \mathrm{d}x_I.$$

For example,

$$\mathrm{d}\left((x^2 + y + z^3)\mathrm{d}x\right) = \frac{\partial(x^2 + y + z^3)}{\partial x}\mathrm{d}x \wedge \mathrm{d}x$$

$$+ \frac{\partial(x^2 + y + z^3)}{\partial y}\mathrm{d}y \wedge \mathrm{d}x$$

$$+ \frac{\partial(x^2 + y + z^3)}{\partial z}\mathrm{d}z \wedge \mathrm{d}x$$

$$= - \mathrm{d}x \wedge \mathrm{d}y - 3z^2\mathrm{d}x \wedge \mathrm{d}z.$$

Theorem 9.1.1. *For any k-form ω, we have*

$$\mathrm{d}(\mathrm{d}\omega) = 0.$$

The proof is a calculation that we leave for the exercises. At a critical stage you must use that the order of differentiation does not matter, meaning that

$$\frac{\partial^2 f}{\partial x_i \partial x_j} = \frac{\partial^2 f}{\partial x_j \partial x_i}.$$

Theorem 9.1.2 (Poincaré's Lemma). *In $\mathbb{R}^n$, let ω be a k-form such that*

$$\mathrm{d}\omega = 0.$$

Then there exists a $(k-1)$-form τ such that

$$\omega = \mathrm{d}\tau.$$

(For a proof, see theorem 4.11 in [61]. It must be emphasized that we are working in $\mathbb{R}^n$. Later we will have differential forms on more complicated spaces. On these, Poincaré's Lemma need not hold.)

Let us look at an example. In $\mathbb{R}^3$, let

$$\omega = f\,dx + g\,dy + h\,dz.$$

Suppose that

$$d\omega = 0.$$

This means that

$$0 = d\omega$$

$$= d(f\,dx + g\,dy + h\,dz)$$

$$= \frac{\partial f}{\partial y}dy \wedge dx + \frac{\partial f}{\partial z}dz \wedge dx$$

$$+ \frac{\partial g}{\partial x}dx \wedge dy + \frac{\partial g}{\partial z}dz \wedge dy$$

$$+ \frac{\partial h}{\partial x}dx \wedge dz + \frac{\partial h}{\partial y}dy \wedge dz$$

$$= \left(\frac{\partial g}{\partial x} - \frac{\partial f}{\partial y}\right)dx \wedge dy + \left(\frac{\partial h}{\partial x} - \frac{\partial f}{\partial z}\right)dx \wedge dz + \left(\frac{\partial h}{\partial y} - \frac{\partial g}{\partial z}\right)dy \wedge dz.$$

Thus $d\omega = 0$ is just a succinct notation for the set of conditions

$$\frac{\partial f}{\partial y} = \frac{\partial g}{\partial x}$$

$$\frac{\partial f}{\partial z} = \frac{\partial h}{\partial x}$$

$$\frac{\partial g}{\partial z} = \frac{\partial h}{\partial y}.$$

The preceding theorem is stating that there is a 0-form τ such that $d\tau = \omega$. Now a 0-form is just a function $\tau(x,y,z)$, which in turn means

$$d\tau = \frac{\partial \tau}{\partial x}dx + \frac{\partial \tau}{\partial y}dy + \frac{\partial \tau}{\partial z}dz.$$

Thus $d\tau = \omega$ means that there is a function τ such that

$$f = \frac{\partial \tau}{\partial x}$$

$$g = \frac{\partial \tau}{\partial y}$$

$$h = \frac{\partial \tau}{\partial z}.$$

We will see in Chapter 11 that this theorem is a general way of stating Theorem 7.2.2 from vector calculus, namely, that the scalar and vector potentials always exist.

9.2. Tools for Measuring

We will now see that differential forms are tools for measuring. A 1-form is a measuring tool for curves, a 2-form a measuring tool for surfaces, and, in general, a k-form is a measuring tool for k-dimensional objects. In general, if M is a k-dimensional subset of $\mathbb{R}^n$ and ω is a k-form, we want to interpret

$$\int_M \omega$$

as a number. We will start with seeing how to use 1-forms to make these measurements for curves in $\mathbb{R}^3$, then turn to the 2-form case for surfaces in $\mathbb{R}^3$, and finally see the general situation. As mentioned earlier, in manipulating forms, we rarely need actually to think about them as these "measuring tools."

9.2.1. Curves in $\mathbb{R}^3$

We will consider curves in $\mathbb{R}^3$ as images of maps

$$\gamma : [a,b] \to \mathbb{R}^3$$

where

$$\gamma(u) = (x(u), y(u), z(u)).$$

We will frequently denote this curve by γ.

For example,

$$\gamma(u) = (u, 3u, 5u)$$

describes a straight line through the point $\gamma(0) = (0,0,0)$ and the point $\gamma(1) = (1,3,5)$. The curve

$$\gamma(u) = (\cos(u), \sin(u), 1)$$

describes a unit circle in the plane $z = 1$.

The tangent vector to a curve $\gamma(u)$ at a point u is the vector

$$\left(\frac{dx(u)}{du}, \frac{dy(u)}{du}, \frac{dz(u)}{du} \right).$$

Now consider a 1-form

$$\omega = f(x,y,z)dx + g(x,y,z)dy + h(x,y,z)dz.$$

Recalling that $\gamma : [a,b] \to \mathbb{R}^3$, we make the following definitions:

$$\int_\gamma f(x,y,z)dx = \int_a^b f(x(u),y(u),z(u))\left(\frac{dx}{du}\right)du$$

$$\int_\gamma g(x,y,z)dy = \int_a^b g(x(u),y(u),z(u))\left(\frac{dy}{du}\right)du$$

$$\int_\gamma h(x,y,z)dz = \int_a^b h(x(u),y(u),z(u))\left(\frac{dz}{du}\right)du.$$

Then we define

$$\int_\gamma \omega = \int_\gamma f\,dx + \int_\gamma g\,dy + \int_\gamma h\,dz.$$

This is called the *path integral*. (In many multivariable calculus classes, it is called the line integral, even though the integration is not necessarily done over a straight line but over a more general curve.)

The definition for $\int_\gamma \omega$ explicitly uses the parameterization of the curve $\gamma(u) = (x(u),y(u),z(u))$. Luckily, the actual value of $\int_\gamma \omega$ depends not on this parameterization but only on the image curve γ in $\mathbb{R}^3$. In fact, all of the preceding can be done for curves in $\mathbb{R}^n$, leading to:

Theorem 9.2.1. *Let $\omega \in \Lambda^1(\mathbb{R}^n)$ and let*

$$\gamma : [a,b] \to \mathbb{R}^n$$

be any curve

$$\gamma(u) = (x_1(u),\dots,x_n(u)).$$

Let

$$\mu : [c,d] \to [a,b]$$

be any map such that $\mu(c) = a$ and $\mu(d) = b$. Then

$$\int_{\gamma \circ \mu} \omega = \int_\gamma \omega.$$

This means that the path integral $\int_\gamma \omega$ is independent of parameterization. You are asked to prove a slightly special case of this in the exercises.

9.2.2. Surfaces in $\mathbb{R}^3$

For us, surfaces in $\mathbb{R}^3$ will be described as the image of a map

$$\gamma : [a_1, a_2] \times [b_1, b_2] \to \mathbb{R}^3,$$

where

$$\gamma(u, v) = (x(u, v), y(u, v), z(u, v)).$$

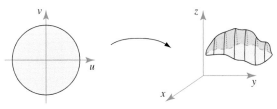

Figure 9.1

The Jacobian of the map γ is

$$D(\gamma) = \begin{pmatrix} \dfrac{\partial x}{\partial u} & \dfrac{\partial x}{\partial v} \\[2ex] \dfrac{\partial y}{\partial u} & \dfrac{\partial y}{\partial v} \\[2ex] \dfrac{\partial z}{\partial u} & \dfrac{\partial z}{\partial v} \end{pmatrix}.$$

From multivariable calculus, we know that at the point $\gamma(a, b) = (x(a, b), y(a, b), z(a, b))$, the tangent plane is spanned by the columns of $D(\gamma)$.
Let

$$\omega = f \, dx \wedge dy + g \, dx \wedge dz + h \, dy \wedge dz.$$

We want to be able to give meaning to the symbol

$$\int_\gamma \omega = \int_\gamma (f \, dx \wedge dy + g \, dx \wedge dz + h \, dy \wedge dz).$$

Start with the following definitions:

$$dx \wedge dy \, (D(\gamma)) = dx \wedge dy \begin{pmatrix} \dfrac{\partial x}{\partial u} & \dfrac{\partial x}{\partial v} \\[2ex] \dfrac{\partial y}{\partial u} & \dfrac{\partial y}{\partial v} \\[2ex] \dfrac{\partial z}{\partial u} & \dfrac{\partial z}{\partial v} \end{pmatrix}$$

$$= \det \begin{pmatrix} \dfrac{\partial x}{\partial u} & \dfrac{\partial x}{\partial v} \\[2ex] \dfrac{\partial y}{\partial u} & \dfrac{\partial y}{\partial v} \end{pmatrix}$$

$$= \frac{\partial x}{\partial u}\frac{\partial y}{\partial v} - \frac{\partial y}{\partial u}\frac{\partial x}{\partial v},$$

$$dx \wedge dz \, (D(\gamma)) = dx \wedge dz \begin{pmatrix} \dfrac{\partial x}{\partial u} & \dfrac{\partial x}{\partial v} \\[2ex] \dfrac{\partial y}{\partial u} & \dfrac{\partial y}{\partial v} \\[2ex] \dfrac{\partial z}{\partial u} & \dfrac{\partial z}{\partial v} \end{pmatrix}$$

$$= \det \begin{pmatrix} \dfrac{\partial x}{\partial u} & \dfrac{\partial x}{\partial v} \\[2ex] \dfrac{\partial z}{\partial u} & \dfrac{\partial z}{\partial v} \end{pmatrix}$$

$$= \frac{\partial x}{\partial u}\frac{\partial z}{\partial v} - \frac{\partial z}{\partial u}\frac{\partial x}{\partial v},$$

$$dy \wedge dz \, (D(\gamma)) = dy \wedge dz \begin{pmatrix} \dfrac{\partial x}{\partial u} & \dfrac{\partial x}{\partial v} \\[2ex] \dfrac{\partial y}{\partial u} & \dfrac{\partial y}{\partial v} \\[2ex] \dfrac{\partial z}{\partial u} & \dfrac{\partial z}{\partial v} \end{pmatrix}$$

$$= \det \begin{pmatrix} \dfrac{\partial y}{\partial u} & \dfrac{\partial y}{\partial v} \\[2ex] \dfrac{\partial z}{\partial u} & \dfrac{\partial z}{\partial v} \end{pmatrix}$$

$$= \frac{\partial y}{\partial u}\frac{\partial z}{\partial v} - \frac{\partial z}{\partial u}\frac{\partial y}{\partial v}.$$

Then define each of the following:

$$\int_\gamma f \, dx \wedge dy = \int_{b_1}^{b_2} \int_{a_1}^{a_2} f(x(u,v), y(u,v), z(u,z)) dx \wedge dy(D(\gamma)) \, du dv,$$

$$\int_\gamma g \, dx \wedge dz = \int_{b_1}^{b_2} \int_{a_1}^{a_2} g(x(u,v), y(u,v), z(u,z)) dx \wedge dz(D(\gamma)) \, du dv,$$

$$\int_\gamma h \, dy \wedge dz = \int_{b_1}^{b_2} \int_{a_1}^{a_2} h(x(u,v), y(u,v), z(u,z)) dy \wedge dz(D(\gamma)) \, du dv.$$

Finally define

$$\int_\gamma \omega = \int_\gamma f \, dx \wedge dy + \int_\gamma g \, dx \wedge dz + \int_\gamma h \, dy \wedge dz.$$

We have reduced $\int_\gamma \omega$ to the calculation of three double integrals in the (u, v)-plane.

Although we have the domain for γ being a rectangle $[a_1, a_2] \times [b_1, b_2]$, any region in the (u, v)-plane can be used.

As with curves, while the definition of $\int_\gamma \omega$ seems to depend on the given parameterization for our surface, the actual computed number is parameterization independent, as seen in

Theorem 9.2.2. *Let* $\omega \in \Lambda^2(\mathbb{R}^n)$. *Let*

$$\gamma : [a_1, b_1] \times [c_1, d_1] \to \mathbb{R}^n$$

be any surface

$$\gamma(u, v) = (x_1(u, v), \dots, x_n(u, v)).$$

Let

$$\mu = (\mu_1(s, t), \mu_2(s, t)) : [a_2, b_2] \times [c_2, d_2] \to [a_1, b_1] \times [c_1, d_1]$$

be any map such that the boundary of $[a_2, b_2] \times [c_2, d_2]$ *maps to the boundary of* $[a_1, b_1] \times [c_1, d_1]$. *Then*

$$\int_{\gamma \circ \mu} \omega = \int_\gamma \omega.$$

A slightly special case of the proof is one of the exercises.

9.2.3. k-manifolds in $\mathbb{R}^n$

A k-space in $\mathbb{R}^n$ will be described as the image of a map

$$\gamma : [a_1, b_1] \times \cdots \times [a_k, b_k] \to \mathbb{R}^n,$$

where
$$\gamma(u_1,\ldots,u_k) = (x_1(u_1,\ldots,u_k),\ldots,x_n(u_1,\ldots,u_k)).$$

The Jacobian of the map γ is the following $n \times k$ matrix:

$$D(\gamma) = \begin{pmatrix} \dfrac{\partial x_1}{\partial u_1} & \cdots & \dfrac{\partial x_1}{\partial u_k} \\ & \vdots & \\ \dfrac{\partial x_n}{\partial u_1} & \cdots & \dfrac{\partial x_n}{\partial u_k} \end{pmatrix}.$$

The image of γ is a manifold if the preceding matrix has rank k at all points, which geometrically means that, at each point of the image, the columns of the Jacobian are linearly independent vectors in $\mathbb{R}^n$ spanning the tangent space.

Let each row of $D(\gamma)$ be denoted by R_i, such that

$$D(\gamma) = \begin{pmatrix} R_1 \\ \vdots \\ R_n \end{pmatrix}.$$

If ω is any k-form, we want to make sense out of the symbol $\int_\gamma \omega$. As before with curves and surfaces, we start by looking at an elementary k-form $dx_{i_1} \wedge \cdots \wedge dx_{i_k}$. The key is that at a point on γ, we define

$$dx_{i_1} \wedge \cdots \wedge dx_{i_k}(D(\gamma)) = \det \begin{pmatrix} R_{i_1} \\ \vdots \\ R_{i_k} \end{pmatrix}.$$

Then we define

$$\int_\gamma f\, dx_{i_1} \wedge \cdots \wedge dx_{i_k}$$

to be

$$\int f(x_1(u_1,\ldots,u_k),\ldots,x_n(u_1,\ldots,u_k))(D(\gamma))\, du_1 \cdots du_k,$$

where we are integrating over $[a_1,b_1] \times \cdots \times [a_k,b_k]$ in $\mathbb{R}^k$.

Then if
$$\omega = \sum f_I dx_I,$$

we define

$$\int_\gamma \omega = \sum \int_\gamma f_I dx_I.$$

As with the special cases of curves and surfaces, this integral is actually independent of parameterization.

(For more on differential forms, see [26] or Hubbard and Hubbard's *Vector Calculus, Linear Algebra, and Differential Forms: A Unified Approach*, [33].)

9.3. Exercises

Exercise 9.3.1. *Write down a basis for the vector space $\Lambda^1(\mathbb{R}^4)$, the vector space $\Lambda^2(\mathbb{R}^4)$, the vector space $\Lambda^3(\mathbb{R}^4)$, and the vector space $\Lambda^4(\mathbb{R}^4)$.*

Exercise 9.3.2. *Write down a basis for the vector space $\Lambda^1(\mathbb{R}^5)$, the vector space $\Lambda^2(\mathbb{R}^5)$, the vector space $\Lambda^3(\mathbb{R}^5)$, and the vector space $\Lambda^4(\mathbb{R}^5)$.*

Exercise 9.3.3. *Let*

$$\omega = (x_1^2 + x_2x_3)\mathrm{d}x_1 + x_2x_3^3\mathrm{d}x_2 \in \Lambda^1(\mathbb{R}^3).$$

Compute $\mathrm{d}\omega$.

Exercise 9.3.4. *Let*

$$\omega = (x_1^2 + x_2x_4)\mathrm{d}x_1 \wedge \mathrm{d}x_2 + x_2x_3^3\mathrm{d}x_2 \wedge \mathrm{d}x_3 + \sin(x_2)\mathrm{d}x_3 \wedge \mathrm{d}x_4 \in \Lambda^2(\mathbb{R}^4).$$

Compute $\mathrm{d}\omega$.

We now begin a series of exercises to show that

$$\mathrm{d}^2\omega = 0.$$

Exercise 9.3.5. *Let $f(x,y,z)$ be any function on $\mathbb{R}^3$. Show that*

$$\mathrm{d}\mathrm{d}f = \mathrm{d}^2 f = 0.$$

Exercise 9.3.6. *Let f be any function on $\mathbb{R}^n$. Show that*

$$\mathrm{d}^2 f = 0.$$

Exercise 9.3.7. *Let f be any function on $\mathbb{R}^n$. Set $\omega = f\mathrm{d}x_1$. Show that*

$$\mathrm{d}^2\omega = 0.$$

Exercise 9.3.8. *Let $\omega = \sum f_i\mathrm{d}x_i$ be any 1-form. Show that*

$$\mathrm{d}^2\omega = 0.$$

Exercise 9.3.9. *Let $\omega = f\mathrm{d}x_1 \wedge \mathrm{d}x_2$ be any 2-form on $\mathbb{R}^n$. Show that*

$$\mathrm{d}^2\omega = 0.$$

Exercise 9.3.10. *Let* $\omega = f\,\mathrm{d}x_1$ *be a k-form on* $\mathbb{R}^n$. *Show that*

$$\mathrm{d}^2\omega = 0.$$

Exercise 9.3.11. *Let* ω *be a k-form on* $\mathbb{R}^n$. *Show that*

$$\mathrm{d}^2\omega = 0.$$

Exercise 9.3.12. *Let* $\omega = (x_1^2 + x_2x_3)\mathrm{d}x_1 + x_2x_3^3\mathrm{d}x_2 \in \Lambda^1(\mathbb{R}^3)$. *For the curve*

$$\gamma : [2,6] \to \mathbb{R}^3,$$

given by

$$\gamma(u) = (u, 2+u, 5u),$$

compute

$$\int_\gamma \omega.$$

Exercise 9.3.13. *Let*

$$\omega = (x_1^2 + x_2x_4)\mathrm{d}x_1 \wedge \mathrm{d}x_2 + x_2x_3^3\mathrm{d}x_2 \wedge \mathrm{d}x_3 + x_2\mathrm{d}x_3 \wedge \mathrm{d}x_4 \in \Lambda^2(\mathbb{R}^4).$$

For the surface

$$\gamma : [0,6] \times [0,3] \to \mathbb{R}^4,$$

given by

$$\gamma(u,v) = (u+v, 2u+3v, 5u+v, u-v),$$

compute

$$\int_\gamma \omega.$$

Exercise 9.3.14. *Let*

$$\omega = x_1x_2x_3^2\mathrm{d}x_1 \wedge \mathrm{d}x_2 \wedge \mathrm{d}x_3 \in \Lambda^3(\mathbb{R}^3).$$

For the box

$$\gamma : [0,1] \times [0,2] \times [0,3] \to \mathbb{R}^3,$$

defined by

$$\gamma(u,v,w) = (u+w, v+3w, u),$$

compute

$$\int_\gamma \omega.$$

Exercise 9.3.15. *Let* $f(x_1,x_2,x_3) = x_1^3x_2x_3^2$. *Set*

$$\omega = \mathrm{d}(f).$$

For the curve

$$\gamma : [0,6] \to \mathbb{R}^3,$$

given by

$$\gamma(u) = \left(u^2, 2 + u^3, 5u\right),$$

compute from the definition

$$\int_\gamma \omega.$$

Show that this is equal to

$$f(\gamma(6)) - f(\gamma(0)).$$

Exercise 9.3.16. *Let* $\omega = (x_1^2 + x_2 x_3)\mathrm{d}x_1 + x_2 x_3^3 \mathrm{d}x_2 \in \Lambda^1(\mathbb{R}^3)$. *Let*

$$\gamma : [2,6] \to \mathbb{R}^3$$

be the curve

$$\gamma(u) = \left(u^2, 2 + u^3, 5u\right).$$

Let

$$\mu : [0,4] \to [2,6]$$

be defined by

$$\mu(t) = t + 2.$$

For the curve $\gamma \circ \mu : [0,4] \to \mathbb{R}^3$, *compute*

$$\int_{\gamma \circ \mu} \omega.$$

Show that this integral equals $\int_\gamma \omega$.

Exercise 9.3.17. *Use the same notation as in the last problem, but now let*

$$\mu : [\sqrt{2}, \sqrt{6}] \to [2,6]$$

be defined by

$$\mu(t) = t^2.$$

Again, for the curve $\gamma \circ \mu : [\sqrt{2}, \sqrt{6}] \to \mathbb{R}^3$, *compute*

$$\int_{\gamma \circ \mu} \omega.$$

Show again that this integral equals $\int_\gamma \omega$.

Exercise 9.3.18. *Let* $\omega \in \Lambda^1(\mathbb{R}^n)$. *Let*

$$\gamma : [a,b] \to \mathbb{R}^n$$

be any curve

$$\gamma(u) = (x_1(u), \ldots, x_n(u)).$$

Let

$$\mu : [c,d] \to [a,b]$$

be any map such that $\mu(c) = a$ and $\mu(d) = b$ and, for points in $[c,d]$, $d\mu/dt > 0$. Show that

$$\int_{\gamma \circ \mu} \omega = \int_{\gamma} \omega.$$

This means that the path integral $\int_{\gamma} \omega$ is independent of parameterization.

Exercise 9.3.19. *Let $\omega \in \Lambda^2(\mathbb{R}^n)$. Let*

$$\gamma : [a_1, b_1] \times [c_1, d_1] \to \mathbb{R}^n$$

be any surface

$$\gamma(u,v) = (x_1(u,v), \ldots, x_n(u,v)).$$

Let

$$\mu = (\mu_1(s,t), \mu_2(s,t)) : [a_2, b_2] \times [c_2, d_2] \to [a_1, b_1] \times [c_1, d_1]$$

be any map such that the boundary of $[a_2, b_2] \times [c_2, d_2]$ maps to the boundary of $[a_1, b_1] \times [c_1, d_1]$ and, for all points in $[a_2, b_2] \times [c_2, d_2]$, the Jacobian is positive, meaning that

$$\det \begin{pmatrix} \dfrac{\partial \mu_1}{\partial s} & \dfrac{\partial \mu_1}{\partial t} \\ \dfrac{\partial \mu_2}{\partial s} & \dfrac{\partial \mu_2}{\partial t} \end{pmatrix} > 0.$$

Show that

$$\int_{\gamma \circ \mu} \omega = \int_{\gamma} \omega.$$

This means that the surface integral $\int_{\gamma} \omega$ is independent of parameterization.

Exercise 9.3.20. *Generalize the last few problems to the integral of*

$$\omega \in \Lambda^k(\mathbb{R}^n)$$

along a k-manifold. Do not do the proof (unless you so desire), but state the appropriate theorem.

10

The Hodge $\star$ Operator

Summary: The goal of this chapter is to define the exterior algebra $\Lambda(\mathbb{R}^n)$ and then to define and understand, in several different contexts, certain natural linear maps

$$\star : \Lambda^k(\mathbb{R}^n) \to \Lambda^{n-k}(\mathbb{R}^n).$$

We will start with the standard Hodge $\star$ operator for the exterior algebra. Using this $\star$ operator, we will then show how the various operations from vector calculus, such as the gradient, the curl, and the divergence, have natural interpretations in the language of the exterior algebra. Then we will generalize the Hodge $\star$ operator by placing different inner products on the vector spaces $\Lambda^k(\mathbb{R}^n)$, allowing us to interpret the Hodge $\star$ operator for the Minkowski metric, which will be critical for the next chapter, where we will interpret Maxwell's equations in terms of differential forms.

10.1. The Exterior Algebra and the $\star$ Operator

For this section, our goal is to define a natural linear map

$$\star : \Lambda^k(\mathbb{R}^n) \to \Lambda^{n-k}(\mathbb{R}^n),$$

called the Hodge $\star$ operator, or, simply, the $\star$ operator. The symbol $\star$ is pronounced "star."

First, there is a natural way to combine a k-form ω with an l-form τ to create a $(k + l)$-form, which we will denote as

$$\omega \wedge \tau.$$

It is best to see this via an example. Suppose we are in $\mathbb{R}^7$. Let

$$\omega = \mathrm{d}x_1 \wedge \mathrm{d}x_3 + \mathrm{d}x_2 \wedge \mathrm{d}x_5$$

be a 2-form and

$$\tau = dx_2 \wedge dx_4 \wedge dx_6 + dx_3 \wedge dx_4 \wedge dx_6$$

be a 3-form. Then $\omega \wedge \tau$ will be the 5-form

$$\omega \wedge \tau = (dx_1 \wedge dx_3 + dx_2 \wedge dx_5) \wedge (dx_2 \wedge dx_4 \wedge dx_6 + dx_3 \wedge dx_4 \wedge dx_6)$$
$$= (dx_1 \wedge dx_3) \wedge (dx_2 \wedge dx_4 \wedge dx_6)$$
$$+ (dx_1 \wedge dx_3) \wedge (dx_3 \wedge dx_4 \wedge dx_6)$$
$$+ (dx_2 \wedge dx_5) \wedge (dx_2 \wedge dx_4 \wedge dx_6)$$
$$+ (dx_2 \wedge dx_5) \wedge (dx_3 \wedge dx_4 \wedge dx_6)$$
$$= -dx_1 \wedge dx_2 \wedge dx_3 \wedge dx_4 \wedge dx_6 + dx_2 \wedge dx_3 \wedge dx_4 \wedge dx_5 \wedge dx_6.$$

We denote the space of all differential forms on $\mathbb{R}^n$ by $\Lambda(\mathbb{R}^n)$, which we call the *exterior algebra*.

Now we turn to the definition of $\star : \Lambda^k(\mathbb{R}^n) \to \Lambda^{n-k}(\mathbb{R}^n)$.

The vector space $\Lambda^n(\mathbb{R}^n)$ of n-forms on $\mathbb{R}^n$ is one dimensional, with basis element

$$dx_1 \wedge dx_2 \wedge \cdots \wedge dx_n.$$

Any k-form ω is the sum of elementary k-forms:

$$\omega = \sum f_I dx_I$$

where, as before, we are summing over all $I = (i_i, \ldots, i_k)$ and where $dx_I = dx_{i_1} \wedge \cdots \wedge dx_{i_k}$. We will first define $\star : \Lambda^k(\mathbb{R}^n) \to \Lambda(\mathbb{R}^{n-k})$ as a map on the elementary k-forms and then simply define

$$\star \omega = \sum f_I \star (dx_I),$$

for a general $k-$form.

So, given an elementary k-form dx_I, we define $\star(dx_I)$ to be the elementary $(n-k)$-form dx_J such that

$$dx_I \wedge dx_J = dx_1 \wedge dx_2 \wedge \cdots \wedge dx_n.$$

Let us look at some examples. We start with $\mathbb{R}^3$, with coordinates (x, y, z). We write a basis for the space of three-forms on $\mathbb{R}^3$ as

$$dx \wedge dy \wedge dz.$$

Then we have

$$\star(dx) = dy \wedge dz$$
$$\star(dy) = -dx \wedge dz$$
$$\star(dz) = dx \wedge dy.$$

The minus sign in the second equation arises because

$$dy \wedge (-dx \wedge dz) = -dy \wedge dx \wedge dz = dx \wedge dy \wedge dz.$$

10.2. Vector Fields and Differential Forms

Our earlier language of vector fields, including gradients, curls, and divergences, can now be recast into the language of differential forms. The key is that there are four linear maps (using the notation in section 6.3 of [26]):

$$
\begin{array}{llll}
T_0 & : & \text{functions on } \mathbb{R}^3 & \rightarrow \quad \text{0-forms} \\
T_1 & : & \text{vector fields on } \mathbb{R}^3 & \rightarrow \quad \text{1-forms} \\
T_2 & : & \text{vector fields on } \mathbb{R}^3 & \rightarrow \quad \text{2-forms} \\
T_3 & : & \text{functions on } \mathbb{R}^3 & \rightarrow \quad \text{3-forms}
\end{array}
$$

with each defined as

$$T_0(f(x,y,z)) = f(x,y,z)$$
$$T_1(F_1(x,y,z), F_2(x,y,z), F_3(x,y,z)) = F_1 dx + F_2 dy + F_3 dz$$
$$T_2(F_1(x,y,z), F_2(x,y,z), F_3(x,y,z)) = \star(F_1 dx + F_2 dy + F_3 dz)$$
$$= F_1 dy \wedge dz - F_2 dx \wedge dz + F_3 dx \wedge dy$$
$$T_3(f(x,y,z)) = f(x,y,z) dx \wedge dy \wedge dz.$$

Then we have that $\nabla(f)$ will correspond to the exterior derivative of a function:

$$\nabla(f) = \left(\frac{\partial f}{\partial x}, \frac{\partial f}{\partial y}, \frac{\partial f}{\partial z}\right)$$
$$\xrightarrow{T_1} \frac{\partial f}{\partial x} dx + \frac{\partial f}{\partial y} dy + \frac{\partial f}{\partial z} dz$$
$$= d(f).$$

Similarly we have:

$$\nabla \times (F_1, F_2, F_3) = \left(\frac{\partial F_3}{\partial y} - \frac{\partial F_2}{\partial z}, \frac{\partial F_1}{\partial z} - \frac{\partial F_3}{\partial x}, \frac{\partial F_2}{\partial x} - \frac{\partial F_1}{\partial y} \right)$$

$$\xrightarrow{T_2} \left(\frac{\partial F_3}{\partial y} - \frac{\partial F_2}{\partial z} \right) dy \wedge dz - \left(\frac{\partial F_1}{\partial z} - \frac{\partial F_3}{\partial x}, \right) dx \wedge dz$$

$$+ \left(\frac{\partial F_2}{\partial x} - \frac{\partial F_1}{\partial y} \right) dx \wedge dy$$

$$= d(F_1 dx + F_2 dy + F_3 dz)$$

$$= d(T_1(F_1, F_2, F_3))$$

$$\nabla \cdot (F_1, F_2, F_3) = \frac{\partial F_1}{\partial x} + \frac{\partial F_2}{\partial y} + \frac{\partial F_3}{\partial z}$$

$$\xrightarrow{T_3} \left(\frac{\partial F_1}{\partial x} + \frac{\partial F_2}{\partial y} + \frac{\partial F_3}{\partial z} \right) dx \wedge dy \wedge dz$$

$$= d(F_1 dy \wedge dz - F_2 dx \wedge dz + F_3 dx \wedge dy)$$

$$= d(T_2(F_1, F_2, F_3)).$$

Of course, this level of abstraction for its own sake is not worthwhile. What is important is that we have put basic terms of vector calculus into a much more general setting.

10.3. The ⋆ Operator and Inner Products

In Section 10.1 we defined the map $\star : \Lambda^k(\mathbb{R}^n) \to \Lambda^{n-k}(\mathbb{R}^n)$. Actually there are many different possible star operators, each depending on a choice of a basis element for the one-dimensional vector space $\Lambda^n(\mathbb{R}^n)$ and on a choice of an inner product on each vector space $\Lambda^k(\mathbb{R}^n)$.

Let ω be a non-zero element of $\Lambda^n(\mathbb{R}^n)$. We declare this to be our basis element. Let $\langle \cdot, \cdot \rangle$ be a fixed inner product on $\Lambda^k(\mathbb{R}^n)$.

Definition 10.3.1. *Given ω and $\langle \cdot, \cdot \rangle$, for any $\alpha \in \Lambda^k(\mathbb{R}^n)$, we define $\star(\alpha)$ to be the $(n-k)$-form such that, for any $\beta \in \Lambda^k(\mathbb{R}^n)$, we have*

$$\beta \wedge (\star \alpha) = \langle \alpha, \beta \rangle \omega.$$

Let us see that this definition is compatible with the earlier one. For $\mathbb{R}^n$ with coordinates $x_1, x_2, \ldots, x_n$, we let our basis n-form be

$$\omega = dx_1 \wedge \cdots \wedge dx_n.$$

For indices $I = (i_1, \ldots, i_k)$ with $1 \leq i_1 < i_2 < \cdots < i_k \leq n$, we know that various dx_I form a basis for $\Lambda^k(\mathbb{R}^n)$. We choose our inner product such that the dx_I are orthonormal:

$$\langle dx_I, dx_J \rangle = \begin{cases} 1 & \text{if } I = J \\ 0 & \text{if } I \neq J \end{cases}.$$

Let us show that

$$\star dx_1 = dx_2 \wedge \cdots \wedge dx_n.$$

First, we need, for $i \neq 1$,

$$dx_1 \wedge \star dx_1 = \langle dx_1, dx_1 \rangle dx_1 \wedge \cdots \wedge dx_n$$
$$= dx_1 \wedge \cdots \wedge dx_n$$
$$dx_i \wedge \star dx_1 = \langle dx_1, dx_i \rangle dx_1 \wedge \cdots \wedge dx_n$$
$$= 0 \cdot dx_1 \wedge \cdots \wedge dx_n$$
$$= 0.$$

There is only one $(n-1)$-form that has these properties. Hence

$$\star dx_1 = dx_2 \wedge \cdots \wedge dx_n.$$

Similarly, we have

$$\star dx_2 = -dx_1 \wedge dx_3 \wedge \cdots \wedge dx_n,$$

since for $i \neq 1$,

$$dx_2 \wedge \star dx_2 = \langle dx_2, dx_2 \rangle dx_1 \wedge \cdots \wedge dx_n$$
$$= dx_2 \wedge (-(dx_1 \wedge dx_3 \cdots \wedge dx_n))$$
$$dx_i \wedge \star dx_1 = \langle dx_1, dx_i \rangle dx_1 \wedge \cdots \wedge dx_n$$
$$= 0 \cdot dx_1 \wedge \cdots \wedge dx_n$$
$$= 0.$$

10.4. Inner Products on $\Lambda(\mathbb{R}^n)$

From the last section, given an inner product on $\Lambda^k(\mathbb{R}^n)$ and a basis n-form on $\Lambda^n(\mathbb{R}^n)$, there is a star operator. But, to a large extent, the space of all exterior forms $\Lambda(\mathbb{R}^n)$ is one object, not a bunch of independent vector spaces $\Lambda^1(\mathbb{R}^n), \Lambda^2(\mathbb{R}^n), \ldots, \Lambda^n(\mathbb{R}^n)$.

In this section we will choose an inner product for $\Lambda^1(\mathbb{R}^n)$ and show how this inner product induces inner products on each of the other vector spaces

$\Lambda^k(\mathbb{R}^n)$. First to recall some linear algebra facts. Let V be a vector space with basis $v_1, v_2, \ldots, v_n$. An inner product $\langle \cdot, \cdot \rangle$ on V is completely determined by knowing the values

$$\langle v_i, v_j \rangle = a_{ij},$$

for all $1 \leq i, j \leq n$. Given two vectors $u = \sum_{i=1}^n \alpha_i v_i$ and $w = \sum_{j=1}^n \beta_j v_j$, the inner product is

$$\langle u, w \rangle = \left\langle \sum_{i=1}^n \alpha_i v_i, \sum_{j=1}^n \beta_j v_j \right\rangle$$

$$= \sum_{i,j=1}^n a_{ij} \alpha_i \beta_j,$$

or, in matrix notation,

$$\langle u, w \rangle = (\alpha_1, \cdots, \alpha_n) \begin{pmatrix} a_{11} & \cdots & a_{1n} \\ & \vdots & \\ a_{n1} & \cdots & a_{nn} \end{pmatrix} \begin{pmatrix} \beta_1 \\ \vdots \\ \beta_n \end{pmatrix}$$

$$= \sum_{i,j=1}^n a_{ij} \alpha_i \beta_j.$$

Now we turn to the exterior algebra $\Lambda(\mathbb{R}^n)$. Our basis for $\Lambda^1(\mathbb{R}^n)$ is $dx_1, dx_2, \ldots, dx_n$. Suppose we have an inner product on $\Lambda^1(\mathbb{R}^n)$. Thus we know the values for $\langle dx_i, dx_j \rangle$. A basis for $\Lambda^k(\mathbb{R}^n)$ is formed from all the $dx_I = dx_{i_1} \wedge \cdots \wedge dx_{i_k}$. Then we define our inner product on $\Lambda^k(\mathbb{R}^n)$ by setting, for $I = (i_1, \ldots, i_k)$ and $J = (j_1, \ldots, j_k)$,

$$\langle dx_I, dx_J \rangle = \sum_{\sigma \in S_k} \text{sign}(\sigma) \langle dx_{i_1}, dx_{j_{\sigma(1)}} \rangle \langle dx_{i_2}, dx_{j_{\sigma(2)}} \rangle \cdots \langle dx_{i_k}, dx_{j_{\sigma(k)}} \rangle,$$

where S_k is the group of all the permutations on k elements.

Of course, we need to look at an example. For 2-forms, this definition gives us

$$\langle dx_{i_1} \wedge dx_{i_2}, dx_{j_1} \wedge dx_{j_2} \rangle = \langle dx_{i_1}, dx_{j_1} \rangle \langle dx_{i_2}, dx_{j_2} \rangle - \langle dx_{i_1}, dx_{j_2} \rangle \langle dx_{i_2}, dx_{j_1} \rangle$$

and for 3-forms, we have that $\langle dx_{i_1} \wedge dx_{i_2} \wedge dx_{i_3}, dx_{j_1} \wedge dx_{j_2} \wedge dx_{j_3} \rangle$ is

$$
\begin{aligned}
& \langle dx_{i_1}, dx_{j_1} \rangle \langle dx_{i_2}, dx_{j_2} \rangle \langle dx_{i_3}, dx_{j_3} \rangle \\
- \quad & \langle dx_{i_1}, dx_{j_2} \rangle \langle dx_{i_2}, dx_{j_1} \rangle \langle dx_{i_3}, dx_{j_3} \rangle \\
- \quad & \langle dx_{i_1}, dx_{j_3} \rangle \langle dx_{i_2}, dx_{j_2} \rangle \langle dx_{i_3}, dx_{j_1} \rangle \\
- \quad & \langle dx_{i_1}, dx_{j_1} \rangle \langle dx_{i_2}, dx_{j_3} \rangle \langle dx_{i_3}, dx_{j_2} \rangle \\
+ \quad & \langle dx_{i_1}, dx_{j_2} \rangle \langle dx_{i_2}, dx_{j_3} \rangle \langle dx_{i_3}, dx_{j_1} \rangle \\
+ \quad & \langle dx_{i_1}, dx_{j_3} \rangle \langle dx_{i_2}, dx_{j_1} \rangle \langle dx_{i_3}, dx_{j_2} \rangle.
\end{aligned}
$$

10.5. The ⋆ Operator with the Minkowski Metric

We turn to $\mathbb{R}^4$, with space coordinates x, y, and z and time coordinate t. Special relativity suggests that the four-dimensional vector space $\Lambda^1(\mathbb{R}^4)$ with ordered basis dt, dx, dy, dz should have an inner product given by

$$
\begin{pmatrix}
1 & 0 & 0 & 0 \\
0 & -1 & 0 & 0 \\
0 & 0 & -1 & 0 \\
0 & 0 & 0 & -1
\end{pmatrix}.
$$

(In our chapter on special relativity, we had a c^2 instead of the 1 in the matrix; here we are following the convention of choosing units to make the speed of light one.) Thus for any 1-forms

$$
\alpha = \alpha_1 dt + \alpha_2 dx + \alpha_3 dy + \alpha_4 dz
$$
$$
\beta = \beta_1 dt + \beta_2 dx + \beta_3 dy + \beta_4 dz,
$$

the inner product will be

$$
\langle \alpha, \beta \rangle = (\alpha_1, \alpha_2, \alpha_3, \alpha_4)
\begin{pmatrix}
1 & 0 & 0 & 0 \\
0 & -1 & 0 & 0 \\
0 & 0 & -1 & 0 \\
0 & 0 & 0 & -1
\end{pmatrix}
\begin{pmatrix}
\beta_1 \\
\beta_2 \\
\beta_3 \\
\beta_4
\end{pmatrix}
$$
$$
= \alpha_1 \beta_1 - \alpha_2 \beta_2 - \alpha_3 \beta_3 - \alpha_4 \beta_4.
$$

The induced inner product on the six-dimensional vector space of 2-forms, with ordered basis

$$
dt \wedge dx, \ dt \wedge dy, \ dt \wedge dz, \ dx \wedge dy, \ dx \wedge dz, \ dy \wedge dz
$$

is

$$\begin{pmatrix} -1 & 0 & 0 & 0 & 0 & 0 \\ 0 & -1 & 0 & 0 & 0 & 0 \\ 0 & 0 & -1 & 0 & 0 & 0 \\ 0 & 0 & 0 & 1 & 0 & 0 \\ 0 & 0 & 0 & 0 & 1 & 0 \\ 0 & 0 & 0 & 0 & 0 & 1. \end{pmatrix}.$$

As an example,

$$\begin{aligned} \langle dt \wedge dx, dt \wedge dx \rangle &= \langle dt, dt \rangle \langle dx, dx \rangle - \langle dt, dx \rangle \langle dx, dt \rangle \\ &= 1 \cdot (-1) - 0 \cdot 0 \\ &= -1, \end{aligned}$$

while

$$\begin{aligned} \langle dt \wedge dx, dt \wedge dy \rangle &= \langle dt, dt \rangle \langle dx, dy \rangle - \langle dt, dy \rangle \langle dx, dt \rangle \\ &= 1 \cdot 0 - 0 \cdot 0 \\ &= 0. \end{aligned}$$

The basis elements are mutually orthogonal, and we have

$$\langle dt \wedge dx, dt \wedge dx \rangle = \langle dt \wedge dy, dt \wedge dy \rangle = \langle dt \wedge dz, dt \wedge dz \rangle = -1$$

and

$$\langle dx \wedge dy, dx \wedge dy \rangle = \langle dy \wedge dz, dy \wedge dz \rangle = \langle dy \wedge dz, dy \wedge dz \rangle = 1.$$

In a similar manner, we have that the induced inner product on the space of 3-forms $\Lambda^3(\mathbb{R}^4)$ will have the basis elements

$$dx \wedge dy \wedge dz, dt \wedge dx \wedge dy, dt \wedge dx \wedge dz, dt \wedge dy \wedge dz$$

being mutually orthogonal and

$$\langle dx \wedge dy \wedge dz, dx \wedge dy \wedge dz \rangle = -1$$
$$\langle dt \wedge dx \wedge dy, dt \wedge dx \wedge dy \rangle = 1$$
$$\langle dt \wedge dx \wedge dz, dt \wedge dx \wedge dz \rangle = 1$$
$$\langle dt \wedge dy \wedge dz, dt \wedge dy \wedge dz \rangle = 1.$$

This allows us now to write down the ⋆ operator with respect to the Minkowski inner product. We choose our basis for $\Lambda^4(\mathbb{R}^4)$ to be $dt \wedge dx \wedge dy \wedge dz$. Then

$$
\begin{aligned}
\star dt &= dx \wedge dy \wedge dz \\
\star dx &= dt \wedge dy \wedge dz \\
\star dy &= -dt \wedge dx \wedge dz \\
\star dz &= dt \wedge dx \wedge dy \\
\star(dt \wedge dx) &= -dy \wedge dz \\
\star(dt \wedge dy) &= dx \wedge dz \\
\star(dt \wedge dz) &= -dx \wedge dy \\
\star(dx \wedge dy) &= dt \wedge dz \\
\star(dx \wedge dz) &= -dt \wedge dy \\
\star(dy \wedge dz) &= dt \wedge dx \\
\star(dx \wedge dy \wedge dz) &= dt \\
\star(dt \wedge dx \wedge dy) &= dz \\
\star(dt \wedge dx \wedge dz) &= -dy \\
\star(dt \wedge dy \wedge dz) &= dx.
\end{aligned}
$$

10.6. Exercises

Exercise 10.6.1. *Using the Hodge $\star$ operator of Section 10.1, find $\star(dx_1)$ and $\star(dx_2)$ for $dx_1, dx_2 \in \Lambda^1(\mathbb{R}^6)$.*

Exercise 10.6.2. *Using the Hodge $\star$ operator of Section 10.1, find*

$$\star(dx_2 \wedge dx_4)$$

and

$$\star(dx_1 \wedge dx_4)$$

for $dx_2 \wedge dx_4, dx_1 \wedge dx_4 \in \Lambda^2(\mathbb{R}^6)$.

Exercise 10.6.3. *Using the Hodge $\star$ operator of Section 10.1, find*

$$\star(dx_2 \wedge dx_3 \wedge dx_4)$$

and

$$\star(dx_1 \wedge dx_3 \wedge dx_4)$$

for $dx_2 \wedge dx_3 \wedge dx_4, dx_1 \wedge dx_3 \wedge dx_4 \in \Lambda^3(\mathbb{R}^6)$.

Exercise 10.6.4. *Using the Hodge $\star$ operator of Section 10.1, find*

$$\star(dx_2 \wedge dx_3 \wedge dx_4 \wedge dx_5)$$

and

$$\star(dx_1 \wedge dx_3 \wedge dx_4 \wedge dx_5)$$

for $dx_2 \wedge dx_3 \wedge dx_4 \wedge dx_5, dx_1 \wedge dx_3 \wedge dx_4 \wedge dx_5 \in \Lambda^4(\mathbb{R}^6).$

Exercise 10.6.5. *Using the Hodge ⋆ operator of Section 10.1, find*

$$\star(dx_2 \wedge dx_3 \wedge dx_4 \wedge dx_5 \wedge dx_6)$$

and

$$\star(dx_1 \wedge dx_3 \wedge dx_4 \wedge dx_5 \wedge dx_6)$$

for $dx_2 \wedge dx_3 \wedge dx_4 \wedge dx_5 \wedge dx_6, dx_1 \wedge dx_3 \wedge dx_4 \wedge dx_5 \wedge dx_6 \in \Lambda^5(\mathbb{R}^6).$

Exercise 10.6.6. *Define an inner product on* $\Lambda^1(\mathbb{R}^3)$ *by setting*

$$\langle dx_1, dx_1 \rangle = 2, \ \langle dx_2, dx_2 \rangle = 4, \ \langle dx_3, dx_3 \rangle = 1$$

and

$$\langle dx_1, dx_2 \rangle = 1, \ \langle dx_2, dx_3 \rangle = 0, \ \langle dx_1, dx_3 \rangle = 0.$$

With the ordering dx_1, dx_2, dx_3, *write this inner product as a* 3×3 *symmetric matrix.*

Exercise 10.6.7. *With the inner product on* $\Lambda^1(\mathbb{R}^3)$ *of the previous problem, find the corresponding inner products on* $\Lambda^2(\mathbb{R}^3)$ *and* $\Lambda^3(\mathbb{R}^3)$. *Using the ordering* $dx_1 \wedge dx_2, dx_1 \wedge dx_3, dx_2 \wedge dx_3$, *write the inner product on* $\Lambda^2(\mathbb{R}^3)$ *as a* 3×3 *symmetric matrix.*

Exercise 10.6.8. *Using the Hodge ⋆ operator as defined in Section 10.3 and the inner product in the previous problem, find*

$$\star(dx_1), \star(dx_2), \star(dx_3),$$

with $dx_1 \wedge dx_2 \wedge dx_3$ *as the basis element of* $\Lambda^3(\mathbb{R}^3).$

Exercise 10.6.9. *Using the Hodge ⋆ operator as defined in Section 10.3 and the inner product in the previous problem, find*

$$\star(dx_1 \wedge dx_2), \star(dx_2 \wedge dx_3), \star(dx_1 \wedge dx_3),$$

with $dx_1 \wedge dx_2 \wedge dx_3$ *as the basis element of* $\Lambda^3(\mathbb{R}^3).$

Exercise 10.6.10. *Define an inner product on* $\Lambda^1(\mathbb{R}^3)$ *by setting*

$$\langle dx_1, dx_1 \rangle = 3, \ \langle dx_2, dx_2 \rangle = 4, \ \langle dx_3, dx_3 \rangle = 5$$

and

$$\langle dx_1, dx_2 \rangle = 1, \ \langle dx_1, dx_3 \rangle = 2, \ \langle dx_2, dx_3 \rangle = 1.$$

With the ordering dx_1, dx_2, dx_3, *write this inner product as a* 3×3 *symmetric matrix.*

Exercise 10.6.11. *With the inner product on $\Lambda^1(\mathbb{R}^3)$ of the previous problem, find the corresponding inner products on $\Lambda^2(\mathbb{R}^3)$ and $\Lambda^3(\mathbb{R}^3)$. Using the ordering $dx_1 \wedge dx_2, dx_1 \wedge dx_3, dx_2 \wedge dx_3$, write the inner product on $\Lambda^2(\mathbb{R}^3)$ as a 3×3 symmetric matrix.*

Exercise 10.6.12. *Using the Hodge $\star$ operator as defined in Sections 10.3 and 10.4 and the inner product in the previous problem, find*

$$\star(dx_1), \star(dx_2), \star(dx_3),$$

with $dx_1 \wedge dx_2 \wedge dx_3$ the basis element of $\Lambda^3(\mathbb{R}^3)$.

Exercise 10.6.13. *Using the Hodge $\star$ operator as defined in 10.3 and 10.4 and the inner product in the previous problem, find*

$$\star(dx_1 \wedge dx_2), \star(dx_2 \wedge dx_3), \star(dx_1 \wedge dx_3),$$

with $dx_1 \wedge dx_2 \wedge dx_3$ the basis element of $\Lambda^3(\mathbb{R}^3)$.

Exercise 10.6.14. *Using the Minkowski metric, show*

$$
\begin{aligned}
\star dt &= dx \wedge dy \wedge dz \\
\star dx &= dt \wedge dy \wedge dz \\
\star dy &= -dt \wedge dx \wedge dz \\
\star dz &= dt \wedge dx \wedge dy \\
\star(dt \wedge dx) &= -dy \wedge dz \\
\star(dt \wedge dy) &= dx \wedge dz \\
\star(dt \wedge dz) &= -dx \wedge dy \\
\star(dx \wedge dy) &= dt \wedge dz \\
\star(dx \wedge dz) &= -dt \wedge dy \\
\star(dy \wedge dz) &= dt \wedge dx \\
\star(dx \wedge dy \wedge dz) &= dt \\
\star(dt \wedge dx \wedge dy) &= dz \\
\star(dt \wedge dx \wedge dz) &= -dy \\
\star(dt \wedge dy \wedge dz) &= dx.
\end{aligned}
$$

11

The Electromagnetic Two-Form

Summary: We recast Maxwell's equations into the language of differential forms. This language not only allows for a deeper understanding of Maxwell's equations, and for eventual generalizations to more abstract manifolds, but will also let us recast Maxwell's equations in terms of the calculus of variations via Lagrangians.

11.1. The Electromagnetic Two-Form

We start with the definitions:

Definition 11.1.1. *Let* $E = (E_1, E_2, E_3)$ *and* $B = (B_1, B_2, B_3)$ *be two vector fields. The associated* electromagnetic two-form *is*

$$F = E_1 dx \wedge dt + E_2 dy \wedge dt + E_3 dz \wedge dt$$
$$+ B_1 dy \wedge dz + B_2 dz \wedge dx + B_3 dx \wedge dy.$$

This two-form is also called the Faraday two-form.

Definition 11.1.2. *Let* $\rho(x, y, z, t)$ *be a function and* (J_1, J_2, J_3) *be a vector field. The associated* current one-form *is*

$$J = \rho dt - J_1 dx - J_2 dy - J_3 dz.$$

11.2. Maxwell's Equations via Forms

So far we have just repackaged the vector fields and functions that make up Maxwell's equations. That this repackaging is at least reasonable can be seen via

Theorem 11.2.1. *Vector fields E, B, and J and function ρ satisfy Maxwell's equations if and only if*

$$dF = 0$$

$$\star d \star F = J.$$

Here the star operator is with respect to the Minkowski metric, with basis element $dt \wedge dx \wedge dy \wedge dz$ for $\Lambda(\mathbb{R}^4)$. The proof is a long, though enjoyable, calculation, which we leave for the exercises. While the proof is not conceptually hard, it should be noted how naturally the language of differential forms can be used to describe Maxwell's equations. This language can be generalized to different, more complicated, areas of both mathematics and physics.

11.3. Potentials

We have rewritten Maxwell's equations in the language of differential forms, via the electromagnetic two-form and the current one-form. The question remains as to how far we can go with this rewriting. The answer is that all of our earlier work can be described via differential forms. In this section we will see how the potential function and the potential vector field can be captured via a single one-form.

Recall from the chapter on potentials that for any fields E and B satisfying Maxwell's equations, there are a function ϕ and a vector field A such that $E(x,y,z,t) = -\nabla\phi - \frac{\partial A(x,y,z,t)}{\partial t}$ and $B(x,y,z,t) = \nabla \times A(x,y,z,t)$. We want to encode these potentials into a single one-form.

Definition 11.3.1. *Given a function ϕ and a vector field A, the associated potential one-form is*

$$A = -\phi dt + A_1 dx + A_2 dy + A_3 dz.$$

By setting $A_0 = -\phi$, we write this as

$$A = A_0 dt + A_1 dx + A_2 dy + A_3 dz.$$

Of course, this definition is only worthwhile because the following is true:

Theorem 11.3.1. *Let E and B be vector fields and let F be the corresponding electromagnetic two-form. Then there are a function ϕ and a vector field A*

such that

$$E(x,y,z,t) = -\nabla\phi - \frac{\partial A(x,y,z,t)}{\partial t}$$
$$B(x,y,z,t) = \nabla \times A(x,y,z,t)$$

if and only if, for the potential one-form A, we have

$$F = dA.$$

The proof is another enjoyable calculation, which we will again leave to the exercises.

In our earlier chapter on potentials, we saw that the potentials for the given vector fields E and B are not unique. This has a natural interpretation for differential forms. The key is that for any form ω, we always have

$$d^2\omega = 0,$$

as proven in the exercises in Chapter 9. In particular, for any function f, we always have

$$d^2 f = 0.$$

Suppose A is a potential one-form, meaning that $F = dA$. Then for any function $f(x,y,z,t)$,

$$A + df$$

will also be a potential one-form, since

$$d(A + df) = dA + d^2 f = F.$$

Thus any potential one-form A can be changed to $A + df$.

11.4. Maxwell's Equations via Lagrangians

In Chapter 8 we described how the path of a particle moving in electric and magnetic fields can be described not only as satisfying the differential equation stemming from $F = ma$, but also as the path that was a critical value of a certain integral, the Lagrangian, allowing us to use the machinery from the calculus of variations. We were taking, however, Maxwell's equations as given and thinking of the path of the particle as the "thing" to be varied.

Maxwell's equations are the system of partial differential equations that describe the electric and magnetic fields. Is it possible to reformulate Maxwell's equations themselves in terms of an extremal value for a Lagrangian? The answer to that is the goal of this section. While we are motivating this work by trying to answer the intellectual challenge of recasting Maxwell's equations into the language of the calculus of variations, it is this variational approach that most easily generalizes Maxwell's equations to the weak force and the strong force and to more purely mathematical contexts.

We start with simply stating the Lagrangian L that will yield Maxwell's equations. It should not be clear how this L was discovered.

Definition 11.4.1. *The* electromagnetic Lagrangian *for an electric field E and a magnetic field B is*

$$L = (\star J) \wedge A + \frac{1}{2}(\star F) \wedge F.$$

This is the most efficient way for writing down L but is not directly useful for showing any link with Maxwell's equations.

Now, J is a one-form. Thus $(\star J)$ is a three-form, meaning that $(\star J) \wedge A$ must be a four-form. Similarly, the Faraday form F is a two-form, meaning that $(\star F)$ is also a two-form. Thus $(\star F) \wedge F$ is also a four-form. We know then that L is a four-form on $\mathbb{R}^4$. Thus we will want to find the critical values for

$$\int L \, dx \, dy \, dz \, dt.$$

(If you want, you can think that we want to find the values that minimize the preceding integral. These minimizers will, of course, be among the critical values.)

Proposition 11.4.1. *The electromagnetic Lagrangian L equals*

$$L = (A_0 \rho + A_1 J_1 + A_2 J_2 + A_3 J_3$$

$$+ \frac{1}{2}(E_1^2 + E_2^2 + E_3^2 - B_1^2 - B_2^2 - B_3^2)) dx \wedge dy \wedge dz \wedge dt.$$

The proof is another pleasant calculation, which we leave to the exercises.

We want somehow to find the critical values of this complicated function L. But we need to see exactly what we are varying, in the calculus of variations, to get critical values. Here the potentials become as important as, if not more important than, the fields E and B. We know that

$$E_1 = -\frac{\partial \phi}{\partial x} - \frac{\partial A_1}{\partial t} = \frac{\partial A_0}{\partial x} - \frac{\partial A_1}{\partial t}$$

$$E_2 = -\frac{\partial \phi}{\partial y} - \frac{\partial A_2}{\partial t} = \frac{\partial A_0}{\partial y} - \frac{\partial A_2}{\partial t}$$

$$E_3 = -\frac{\partial \phi}{\partial z} - \frac{\partial A_3}{\partial t} = \frac{\partial A_0}{\partial z} - \frac{\partial A_3}{\partial t}$$

$$B_1 = \frac{\partial A_3}{\partial y} - \frac{\partial A_2}{\partial z}$$

$$B_2 = \frac{\partial A_1}{\partial z} - \frac{\partial A_3}{\partial x}$$

$$B_3 = \frac{\partial A_2}{\partial x} - \frac{\partial A_1}{\partial y}.$$

Thus we can consider our electromagnetic Lagrangian as a function

$$L\left(A_0, A_1, A_2, A_3, \frac{\partial A_0}{\partial t}, \frac{\partial A_0}{\partial x}, \frac{\partial A_0}{\partial y}, \frac{\partial A_0}{\partial z}, \cdots, \frac{\partial A_3}{\partial t}, \frac{\partial A_3}{\partial x}, \frac{\partial A_3}{\partial y}, \frac{\partial A_3}{\partial z}\right).$$

It is the functions A_0, A_1, A_2, A_3 that are varied in L, while the charge function ρ and the current field $J = (J_1, J_2, J_3)$ are givens and fixed. Then we have

Theorem 11.4.1. *Given a charge function ρ and a current field $J = (J_1, J_2, J_3)$, the functions A_0, A_1, A_2, A_3 are the potentials for fields E and B that satisfy Maxwell's equations if and only if they are critical points for*

$$\int L\,dx\,dy\,dz\,dt.$$

Sketch of Proof: We will discuss in the next section that A_0, A_1, A_2, A_3 are critical points for $\int L\,dx\,dy\,dz\,dt$ if and only if they satisfy the Euler-Lagrange equations:

$$\frac{\partial L}{\partial A_0} = \frac{\partial}{\partial t}\left(\frac{\partial L}{\partial(\frac{\partial A_0}{\partial t})}\right) + \frac{\partial}{\partial x}\left(\frac{\partial L}{\partial(\frac{\partial A_0}{\partial x})}\right) + \frac{\partial}{\partial y}\left(\frac{\partial L}{\partial(\frac{\partial A_0}{\partial y})}\right) + \frac{\partial}{\partial z}\left(\frac{\partial L}{\partial(\frac{\partial A_0}{\partial z})}\right)$$

$$\frac{\partial L}{\partial A_1} = \frac{\partial}{\partial t}\left(\frac{\partial L}{\partial(\frac{\partial A_1}{\partial t})}\right) + \frac{\partial}{\partial x}\left(\frac{\partial L}{\partial(\frac{\partial A_1}{\partial x})}\right) + \frac{\partial}{\partial y}\left(\frac{\partial L}{\partial(\frac{\partial A_1}{\partial y})}\right) + \frac{\partial}{\partial z}\left(\frac{\partial L}{\partial(\frac{\partial A_1}{\partial z})}\right)$$

$$\frac{\partial L}{\partial A_2} = \frac{\partial}{\partial t}\left(\frac{\partial L}{\partial(\frac{\partial A_2}{\partial t})}\right) + \frac{\partial}{\partial x}\left(\frac{\partial L}{\partial(\frac{\partial A_2}{\partial x})}\right) + \frac{\partial}{\partial y}\left(\frac{\partial L}{\partial(\frac{\partial A_2}{\partial y})}\right) + \frac{\partial}{\partial z}\left(\frac{\partial L}{\partial(\frac{\partial A_2}{\partial z})}\right)$$

$$\frac{\partial L}{\partial A_3} = \frac{\partial}{\partial t}\left(\frac{\partial L}{\partial(\frac{\partial A_3}{\partial t})}\right) + \frac{\partial}{\partial x}\left(\frac{\partial L}{\partial(\frac{\partial A_3}{\partial x})}\right) + \frac{\partial}{\partial y}\left(\frac{\partial L}{\partial(\frac{\partial A_3}{\partial y})}\right) + \frac{\partial}{\partial z}\left(\frac{\partial L}{\partial(\frac{\partial A_3}{\partial z})}\right).$$

Thus we need the preceding to be equivalent to Maxwell's equations. We will only show here that the third equation in the preceding is equivalent to the part of Maxwell's equations that gives the second coordinate in

$$\nabla \times B = \frac{\partial E}{\partial t} + J.$$

In other words,

$$\frac{\partial B_1}{\partial z} - \frac{\partial B_3}{\partial x} = \frac{\partial E_2}{\partial t} + J_2,$$

which, in the language of the potentials, is equivalent to

$$\frac{\partial^2 A_3}{\partial y \partial z} - \frac{\partial^2 A_2}{\partial z^2} - \frac{\partial^2 A_2}{\partial x^2} + \frac{\partial^2 A_1}{\partial x \partial y} = \frac{\partial^2 A_0}{\partial y \partial t} - \frac{\partial^2 A_2}{\partial t^2} + J_2.$$

The rest of the proof is left for the exercises.

Start with $\partial L/\partial A_2$. Here we must treat A_2 as an independent variable in the function L. The only term in L that contains an A_2 is $A_2 J_2$, which means that

$$\frac{\partial L}{\partial A_2} = J_2.$$

Now to calculate

$$\frac{\partial}{\partial t}\left(\frac{\partial L}{\partial(\frac{\partial A_2}{\partial t})}\right).$$

We start with

$$\frac{\partial L}{\partial(\frac{\partial A_2}{\partial t})}.$$

Here we must treat $\partial A_2/\partial t$ as the independent variable. The only term of L that contains $\partial A_2/\partial t$ is

$$\frac{1}{2} E_2^2 = \frac{1}{2}\left(\frac{\partial A_0}{\partial y} - \frac{\partial A_2}{\partial t}\right)^2.$$

Then

$$\frac{\partial L}{\partial(\frac{\partial A_2}{\partial t})} = -\frac{\partial A_0}{\partial y} + \frac{\partial A_2}{\partial t},$$

giving us

$$\frac{\partial}{\partial t}\left(\frac{\partial L}{\partial(\frac{\partial A_2}{\partial t})}\right) = -\frac{\partial^2 A_0}{\partial y \partial t} + \frac{\partial^2 A_2}{\partial t^2}.$$

Next for $\partial(\partial L/\partial(\partial A_2/\partial x))/\partial x$, where here $\partial A_2/\partial x$ is treated as an independent variable. The only term of L that contains $\partial A_2/\partial x$ is

$$-\frac{1}{2} B_3^2 = -\frac{1}{2}\left(\frac{\partial A_2}{\partial x} - \frac{\partial A_1}{\partial y}\right)^2.$$

Then

$$\frac{\partial L}{\partial(\frac{\partial A_2}{\partial x})} = -\frac{\partial A_2}{\partial x} + \frac{\partial A_1}{\partial y}.$$

Thus

$$\frac{\partial}{\partial x}\left(\frac{\partial L}{\partial(\frac{\partial A_2}{\partial x})}\right) = -\frac{\partial^2 A_2}{\partial x^2} + \frac{\partial^2 A_1}{\partial x \partial y}.$$

The next term, $\partial(\partial L/\partial(\partial A_2/\partial y))/\partial y$, is particularly simple, since L contains no term $\partial A_2/\partial y$, giving us that

$$\frac{\partial}{\partial y}\left(\frac{\partial L}{\partial(\frac{\partial A_2}{\partial y})}\right) = 0.$$

The final term can be calculated, in a similar way, to be

$$\frac{\partial}{\partial z}\left(\frac{\partial L}{\partial(\frac{\partial A_2}{\partial z})}\right) = -\left(\frac{\partial^2 A_3}{\partial y \partial z} - \frac{\partial^2 A_2}{\partial z^2}\right).$$

Putting all of this together, we see that the third of the Euler-Lagrange equations is the same as the second coordinate in $\nabla \times B = \frac{\partial E}{\partial t} + J$, as desired.

The other parts of the Euler-Lagrange equations will imply, by similar means (given in the exercises), the other two coordinates of $\nabla \times B = \frac{\partial E}{\partial t} + J$ and $\nabla \cdot E = \rho$. The remaining two parts of Maxwell's equations ($\nabla \times E = -\partial B/\partial t$ and $\nabla \cdot B = 0$) are built into the machinery of the potential one-form and are thus automatically satisfied. Thus we have a Lagrangian approach to Maxwell's equations.

11.5. Euler-Lagrange Equations for the Electromagnetic Lagrangian

In the last section, in our derivation of Maxwell's equations via Lagrangians, we simply stated the appropriate Euler-Lagrange equations. We now want to give an argument that justifies these equations. Starting with a function L, depending on the four functions $A_0(x,y,z,t)$, $A_1(x,y,z,t)$, $A_2(x,y,z,t)$, $A_3(x,y,z,t)$ and all their partial derivatives with respect to x, y, z and t, the Euler-Lagrange equations are a system of partial differential equations whose solutions are critical points of the integral $\int L\, dx dy dz dt$. We will show how this is done by showing that the critical points satisfy the first Euler-Lagrange equation

$$\frac{\partial L}{\partial A_0} = \frac{\partial}{\partial t}\left(\frac{\partial L}{\partial(\frac{\partial A_0}{\partial t})}\right) + \frac{\partial}{\partial x}\left(\frac{\partial L}{\partial(\frac{\partial A_0}{\partial x})}\right) + \frac{\partial}{\partial y}\left(\frac{\partial L}{\partial(\frac{\partial A_0}{\partial y})}\right) + \frac{\partial}{\partial z}\left(\frac{\partial L}{\partial(\frac{\partial A_0}{\partial z})}\right).$$

Here is the approach:

We assume that we have functions $A_0(x,y,z,t)$, $A_1(x,y,z,t)$, $A_2(x,y,z,t)$, and $A_3(x,y,z,t)$ that are critical values of $\int L\, dx dy dz dt$. Let $\eta(x,y,z,t)$ be any function that is zero on the boundary of what we are integrating over and let ϵ be a number. We perturb the function $A_0(x,y,z,t)$ by setting

$$A_\epsilon = A_0 + \epsilon\eta.$$

Set

$$S(\epsilon) = \int L\left(A_\epsilon, \frac{\partial A_\epsilon}{\partial x}, \frac{\partial A_\epsilon}{\partial y}, \frac{\partial A_\epsilon}{\partial z}, \frac{\partial A_\epsilon}{\partial t}\right) dxdydzdt,$$

where we are suppressing the other variables of L. By assumption, the function $S(\epsilon)$ has a critical value at $\epsilon = 0$. Then we have

$$\frac{dS}{d\epsilon} = 0$$

at $\epsilon = 0$.

Now we have

$$\frac{dS}{d\epsilon} = \frac{d}{d\epsilon}\left(\int L\,dxdydzdt\right)$$

$$= \int \frac{d}{d\epsilon}(L)\,dxdydzdt.$$

Note that to be absolutely rigorous, we would have to justify the interchanging of the derivative and the integral in the preceding expression. To calculate the derivative $dL/d\epsilon$, we use the chain rule to get

$$\frac{dL}{d\epsilon} = \frac{\partial L}{\partial A_0}\frac{dA_\epsilon}{d\epsilon} + \left(\frac{\partial L}{\partial\left(\frac{\partial A_0}{\partial t}\right)}\right)\left(\frac{d\left(\frac{\partial A_\epsilon}{\partial t}\right)}{d\epsilon}\right) + \left(\frac{\partial L}{\partial\left(\frac{\partial A_0}{\partial x}\right)}\right)\left(\frac{d\left(\frac{\partial A_\epsilon}{\partial x}\right)}{d\epsilon}\right)$$

$$+ \left(\frac{\partial L}{\partial\left(\frac{\partial A_0}{\partial y}\right)}\right)\left(\frac{d\left(\frac{\partial A_\epsilon}{\partial y}\right)}{d\epsilon}\right) + \left(\frac{\partial L}{\partial\left(\frac{\partial A_0}{\partial z}\right)}\right)\left(\frac{d\left(\frac{\partial A_\epsilon}{\partial z}\right)}{d\epsilon}\right)$$

$$= \left(\frac{\partial L}{\partial A_0}\right)\eta + \left(\frac{\partial L}{\partial\left(\frac{\partial A_0}{\partial t}\right)}\right)\left(\frac{\partial\eta}{\partial t}\right) + \left(\frac{\partial L}{\partial\left(\frac{\partial A_0}{\partial x}\right)}\right)\left(\frac{\partial\eta}{\partial x}\right)$$

$$+ \left(\frac{\partial L}{\partial\left(\frac{\partial A_0}{\partial y}\right)}\right)\left(\frac{\partial\eta}{\partial y}\right) + \left(\frac{\partial L}{\partial\left(\frac{\partial A_0}{\partial z}\right)}\right)\left(\frac{\partial\eta}{\partial z}\right).$$

In order to put the preceding mess into a more manageable form, we will now use that η is assumed to be zero on the boundary of our region of integration and use integration by parts to show, for any function f, that

$$\int f \frac{\partial \eta}{\partial t} \, dx dy dz dt = - \int \eta \frac{\partial}{\partial t} (f) \, dx dy dz dt.$$

Here, the order of integration does not matter, meaning that

$$\int f \frac{\partial \eta}{\partial t} \, dx dy dz dt = \int \int \int \int f \frac{\partial \eta}{\partial t} \, dt dx dy dz.$$

Here we are now writing out all four integral signs, on the right-hand side; this will make the next few steps easier to follow.

We concentrate on the first integral. Using integration by parts and the fact that η is zero on the boundary, we get that

$$\int \int \int \int f \frac{\partial \eta}{\partial t} \, dt dx dy dz = - \int \int \int \int \eta \frac{\partial}{\partial t} (f) \, dt dx dy dz,$$

giving us what we want.

By a similar argument, we have

$$\int f \frac{\partial \eta}{\partial x} \, dx dy dz dt = - \int \eta \frac{\partial}{\partial x} (f) \, dx dy dz dt$$

$$\int f \frac{\partial \eta}{\partial y} \, dx dy dz dt = - \int \eta \frac{\partial}{\partial y} (f) \, dx dy dz dt$$

$$\int f \frac{\partial \eta}{\partial z} \, dx dy dz dt = - \int \eta \frac{\partial}{\partial z} (f) \, dx dy dz dt.$$

Thus

$$\frac{dL}{d\epsilon} = \eta \left(\frac{\partial L}{\partial A_0} - \frac{\partial}{\partial t} \left(\frac{\partial L}{\partial \left(\frac{\partial A_0}{\partial t} \right)} \right) - \frac{\partial}{\partial x} \left(\frac{\partial L}{\partial \left(\frac{\partial A_0}{\partial x} \right)} \right) \right.$$
$$\left. - \frac{\partial}{\partial y} \left(\frac{\partial L}{\partial \left(\frac{\partial A_0}{\partial y} \right)} \right) - \frac{\partial}{\partial z} \left(\frac{\partial L}{\partial \left(\frac{\partial A_0}{\partial z} \right)} \right) \right).$$

We now have

$$0 = \frac{dS}{d\epsilon}$$

$$= \int \left[\eta \left(\frac{\partial L}{\partial A_0} - \frac{\partial}{\partial t} \left(\frac{\partial L}{\partial \left(\frac{\partial A_0}{\partial t} \right)} \right) - \frac{\partial}{\partial x} \left(\frac{\partial L}{\partial \left(\frac{\partial A_0}{\partial x} \right)} \right) \right. \right.$$
$$\left. \left. - \frac{\partial}{\partial y} \left(\frac{\partial L}{\partial \left(\frac{\partial A_0}{\partial y} \right)} \right) - \frac{\partial}{\partial z} \left(\frac{\partial L}{\partial \left(\frac{\partial A_0}{\partial z} \right)} \right) \right) \right] dx dy dz dt.$$

Since η can be any function that is zero on the boundary, we must have

$$0 = \frac{\partial L}{\partial A_0} - \frac{\partial}{\partial t}\left(\frac{\partial L}{\partial(\frac{\partial A_0}{\partial t})}\right) - \frac{\partial}{\partial x}\left(\frac{\partial L}{\partial(\frac{\partial A_0}{\partial x})}\right) - \frac{\partial}{\partial y}\left(\frac{\partial L}{\partial(\frac{\partial A_0}{\partial y})}\right) - \frac{\partial}{\partial z}\left(\frac{\partial L}{\partial(\frac{\partial A_0}{\partial z})}\right),$$

which yields the first Euler-Lagrange equation. The other three are derived similarly.

For more on this style of approach to Maxwell's equation, I encourage the reader to look at Gross and Kotiuga's *Electromagnetic Theory and Computation: A Topological Approach* [29].

11.6. Exercises

Throughout these exercises, use the star operator with respect to the Minkowski metric, with basis element $dt \wedge dx \wedge dy \wedge dz$ for $\Lambda(\mathbb{R}^4)$.

Exercise 11.6.1. *Prove that the vector fields E, B, and J and function ρ satisfy Maxwell's equations if and only if*

$$dF = 0$$

$$\star d \star F = J.$$

Exercise 11.6.2. *Let E and B be vector fields and let F be the corresponding electromagnetic two-form. Show that there are a function ϕ and a vector field A such that*

$$E(x,y,z,t) = -\nabla\phi - \frac{\partial A(x,y,z,t)}{\partial t}$$

$$B(x,y,z,t) = \nabla \times A(x,y,z,t)$$

if and only if there is a potential one-form A such that

$$F = dA.$$

Exercise 11.6.3. *If F is the electromagnetic two-form, show that*

$$\star F = E_1 dy \wedge dz - E_2 dx \wedge dz + E_3 dx \wedge dy$$
$$- B_1 dx \wedge dt - B_2 dy \wedge dt - B_3 dz \wedge dt.$$

Exercise 11.6.4. *If A is the potential one-form, show that*

$$\star A = -\phi dx \wedge dy \wedge dz + A_1 dy \wedge dz \wedge dt$$
$$- A_2 dx \wedge dz \wedge dt + A_3 dx \wedge dy \wedge dt$$
$$= A_0 dx \wedge dy \wedge dz + A_1 dy \wedge dz \wedge dt$$
$$- A_2 dx \wedge dz \wedge dt + A_3 dx \wedge dy \wedge dt.$$

Exercise 11.6.5. *Letting*

$$L = (\star J) \wedge A + \frac{1}{2}(\star F) \wedge F,$$

show that

$$L = (A_0 \rho + A_1 J_1 + A_2 J_2 + A_3 J_3$$
$$+ 1/2(E_1^2 + E_2^2 + E_3^2 - B_1^2 - B_2^2 - B_3^2))dx \wedge dy \wedge dz \wedge dt.$$

Exercise 11.6.6. *Show that the first of the Euler-Lagrange equations for the electromagnetic Lagrangian*

$$\frac{\partial L}{\partial A_0} = \frac{\partial}{\partial t}\left(\frac{\partial L}{\partial\left(\frac{\partial A_0}{\partial t}\right)}\right) + \frac{\partial}{\partial x}\left(\frac{\partial L}{\partial\left(\frac{\partial A_0}{\partial x}\right)}\right) + \frac{\partial}{\partial y}\left(\frac{\partial L}{\partial\left(\frac{\partial A_0}{\partial y}\right)}\right) + \frac{\partial}{\partial z}\left(\frac{\partial L}{\partial\left(\frac{\partial A_0}{\partial z}\right)}\right)$$

is equivalent to

$$\nabla \cdot E = \rho,$$

the first of Maxwell's equations.

Exercise 11.6.7. *Show that the second of the Euler-Lagrange equations for the electromagnetic Lagrangian*

$$\frac{\partial L}{\partial A_1} = \frac{\partial}{\partial t}\left(\frac{\partial L}{\partial\left(\frac{\partial A_1}{\partial t}\right)}\right) + \frac{\partial}{\partial x}\left(\frac{\partial L}{\partial\left(\frac{\partial A_1}{\partial x}\right)}\right) + \frac{\partial}{\partial y}\left(\frac{\partial L}{\partial\left(\frac{\partial A_1}{\partial y}\right)}\right) + \frac{\partial}{\partial z}\left(\frac{\partial L}{\partial\left(\frac{\partial A_1}{\partial z}\right)}\right)$$

implies the first coordinate in

$$\nabla \times B = \frac{\partial E}{\partial t} + J,$$

the fourth of Maxwell's equations.

Exercise 11.6.8. *Show that the fourth of the Euler-Lagrange equations for the electromagnetic Lagrangian:*

$$\frac{\partial L}{\partial A_3} = \frac{\partial}{\partial t}\left(\frac{\partial L}{\partial\left(\frac{\partial A_3}{\partial t}\right)}\right) + \frac{\partial}{\partial x}\left(\frac{\partial L}{\partial\left(\frac{\partial A_3}{\partial x}\right)}\right) + \frac{\partial}{\partial y}\left(\frac{\partial L}{\partial\left(\frac{\partial A_3}{\partial y}\right)}\right) + \frac{\partial}{\partial z}\left(\frac{\partial L}{\partial\left(\frac{\partial A_3}{\partial z}\right)}\right)$$

implies the third coordinate in

$$\nabla \times B = \frac{\partial E}{\partial t} + J,$$

the fourth of Maxwell's equations.

Exercise 11.6.9. *Let L be the electromagnetic Lagrangian. Show that if $A_0, A_1, A_2,$ and A_3 are critical values for $\int L\,dx\,dy\,dz\,dt$, then they satisfy the second Euler-Lagrange equation*

$$\frac{\partial L}{\partial A_1} = \frac{\partial}{\partial t}\left(\frac{\partial L}{\partial\left(\frac{\partial A_1}{\partial t}\right)}\right) + \frac{\partial}{\partial x}\left(\frac{\partial L}{\partial\left(\frac{\partial A_1}{\partial x}\right)}\right) + \frac{\partial}{\partial y}\left(\frac{\partial L}{\partial\left(\frac{\partial A_1}{\partial y}\right)}\right) + \frac{\partial}{\partial z}\left(\frac{\partial L}{\partial\left(\frac{\partial A_1}{\partial z}\right)}\right).$$

12

Some Mathematics Needed for
Quantum Mechanics

Summary: The goal for this chapter is to define Hilbert spaces and Hermitian operators, both of which will be critical when we try to understand the photoelectric effect of light, which will be explained via the process for quantizing Maxwell's equations over the next few chapters. Thus this chapter is necessary both as a preliminary for the next three chapters on quantum mechanics and as a break from the first part of the book.

12.1. Hilbert Spaces

Hilbert spaces are built into the basic assumptions of quantum mechanics. This section provides a quick overview of Hilbert spaces. The next section will provide a similar overview of Hermitian operators. We will start with the definition for a Hilbert space and then spend time unraveling the definition.

Definition 12.1.1. *A Hilbert space is a complex vector space $\mathcal{H}$ with an inner product that is complete.*

Now we need to define inner product and completeness.

Definition 12.1.2. *An* inner product *on a complex vector space V is a map*

$$\langle \cdot, \cdot \rangle : V \times V \to \mathbb{C}$$

such that

1. *For all vectors $v \in V$, $\langle v, v \rangle$ is a nonnegative real number, with $\langle v, v \rangle = 0$ only if $v = 0$.*
2. *For all vectors $v, w \in V$, we have*

$$\langle v, w \rangle = \overline{\langle w, v \rangle}.$$

3. *For all vectors $u, v, w \in V$ and complex numbers λ and μ, we have*

$$\langle u, \lambda v + \mu w \rangle = \lambda \langle u, v \rangle + \mu \langle u, w \rangle.$$

(Recall that $\overline{a+bi} = a - bi$ is the complex conjugate of the complex number $a + bi$, where both a and b are real numbers.) Inner products are generalizations of the dot product on $\mathbb{R}^2$ from multivariable calculus. Similarly to the dot product, where, recall, for vectors $v, w \in \mathbb{R}^2$, we have

$$v \cdot w = (\text{length of } v)(\text{length of } w)\cos(\theta),$$

with θ being the angle between v and w, an inner product can be used to define lengths and angles on a complex vector space, as follows.

Definition 12.1.3. *Let V be a complex vector space with an inner product $\langle \cdot, \cdot \rangle$. The* length *of a vector $v \in V$ is*

$$|v| = \sqrt{\langle v, v \rangle}.$$

Two vectors v, w are said to be orthogonal *if*

$$\langle v, w \rangle = 0.$$

We will look at some examples in a moment, but we need first to understand what "complete" means. (Our meaning of completeness is the same as the one used in real analysis.) Let V be a complex vector space with an inner product $\langle \cdot, \cdot \rangle$. The inner product can be used to measure distance between vectors $v, w \in V$ by setting

Distance between v and $w = |v - w| = \sqrt{\langle v - w, v - w \rangle}.$

Definition 12.1.4. *A sequence of vectors v_k is a* Cauchy sequence *if, for any $\epsilon > 0$, there exists a positive integer N such that, for all $n, m > N$, we have*

$$|v_n - v_m| < \epsilon.$$

Thus a sequence is Cauchy if the distances between the v_n and v_m can be made arbitrarily small when the n and m are made sufficiently large.

We can now define what it means for a vector space to be complete. (We will then look at an example.)

Definition 12.1.5. *An inner product space V will be* complete *if all Cauchy sequences converge. Hence, if v_k is a Cauchy sequence, there is a vector $v \in V$ such that*

$$\lim_{k \to \infty} v_k = v.$$

The most basic example is the complex vector space $\mathbb{C}$ of dimension one, where the vectors are just complex numbers and the inner product is simply $\langle z, w \rangle = \overline{z}w$. If instead of considering complex vector spaces, we looked at real vector spaces, then the most basic example of a complete space is the real vector space $\mathbb{R}$ of dimension one, where the vectors now are real numbers and the inner product is just multiplication. One of the key properties of the real numbers is that all Cauchy sequences converge. (This is proven in many real analysis texts, such as in chapter 1 of [56] or in [24].)

Let us look at a more complicated example. The key behind this example will be the fact that the series $\sum_{k=1}^{\infty} \frac{1}{k^2}$ converges, and hence that

$$\sum_{k=N}^{\infty} \frac{1}{k^2}$$

can be made arbitrarily small by choosing N large enough. Let V be the vector space consisting of infinite sequences of complex numbers, with all but a finite number being zero. We denote a sequence $(a_1, a_2, a_3, \dots)$ by (a_n). Thus $(a_n) \in V$ if there is an N such that $a_n = 0$ for all $n > N$. Define an inner product on V by setting

$$\langle (a_n), (b_m) \rangle = \sum_{k=1}^{\infty} \bar{a}_k b_k.$$

Though looking like an infinite series, the preceding sum is actually always finite, since only a finite number of the a_k and b_k are non-zero. We want to show that this vector space is *not* complete. Define the sequence, for all n,

$$v_n = \left(1, \frac{1}{2}, \frac{1}{3}, \dots, \frac{1}{n}, 0, \dots \right)$$

For $n < m$, we have that

$$\langle v_m - v_n, v_m - v_n \rangle = \sum_{k=n+1}^{m} \frac{1}{k^2},$$

which can be made arbitrarily small by choosing n large enough. Thus the sequence (v_k) is Cauchy. But this sequence wants to converge to the "vector"

$$v = (1, \frac{1}{2}, \frac{1}{3}, \dots, \frac{1}{n}, \dots),$$

which is not in the vector space V.

This does suggest the following for a Hilbert space.

Theorem 12.1.1. *Define*

$$l_2 = \left\{ (a_1, a_2, a_3, \dots) : a_k \in \mathbb{C} \text{ and } \sum_{k=1}^{\infty} |a_k|^2 < \infty \right\}.$$

For $(a_k),(b_k) \in l_2$, define

$$\langle(a_k),(b_k)\rangle = \sum_{k=1}^{\infty} \overline{a}_k b_k.$$

With this inner product, l_2 is a Hilbert space.

People call this Hilbert space "little ell 2." There are, as to be expected, other vector spaces "little ell p," defined by setting

$$l_p = \{(a_k) : a_k \in \mathbb{C} \text{ and } \sum_{k=1}^{\infty} |a_k|^p < \infty\}.$$

These other spaces do not have an inner product and hence cannot be Hilbert spaces.

Proof. We first need to show that $\langle(a_k),(b_k)\rangle$ is a finite number and hence must show that $\sum_{k=1}^{\infty} \overline{a}_k b_k$ converges whenever

$$\sum_{k=1}^{\infty} |a_k|^2, \sum_{k=1}^{\infty} |b_k|^2 < \infty.$$

Now

$$\left| \sum_{k=1}^{\infty} \overline{a}_k b_k \right| \le \sum_{k=1}^{\infty} |\overline{a}_k b_k| = \sum_{k=1}^{\infty} |a_k||b_k|.$$

The key is that for all complex numbers,

$$|ab| \le |a|^2 + |b|^2.$$

Then

$$\sum_{k=1}^{\infty} |a_k||b_k| \le \sum_{k=1}^{\infty} (|a_k|^2 + |b_k|^2) = \sum_{k=1}^{\infty} |a_k|^2 + \sum_{k=1}^{\infty} |b_k|^2 < \infty.$$

One of the exercises is to show that l_2 is a vector space.

We now must show that all Cauchy sequences converge. (While this proof is standard, we are following section 1.8 in [31].) Let $v_k \in l_2$, with

$$v_k = (a_1(k), a_2(k), a_3(k), \dots),$$

forming a Cauchy sequence. We must show that the sequence of vectors $v_1, v_2, v_3, \dots$ converges to a vector $v = (a_1, a_2, a_3, \dots)$. We know, for $n < m$, that

$$\lim_{n,m \to \infty} |v_m - v_n|^2 = 0.$$

Thus, for any fixed positive integer N we can make

$$\sum_{i=1}^{N} |a_i(m) - a_i(n)|^2$$

arbitrarily small. More precisely, given any $\epsilon > 0$, we can find a positive integer N such that

$$\sum_{k=1}^{\infty} |a_k(m) - a_k(n)|^2 < \epsilon$$

for any $m, n > N$. In particular, for any fixed i, we can make $|a_i(m) - a_i(n)|^2$ arbitrarily small. But this means that the real numbers $a_i(1), a_i(2), a_i(3), \ldots$ form a Cauchy sequence, which must in turn converge to a real number, denoted by a_i. We claim that $(a_1, a_2, a_3, \ldots)$ is our desired sequence.

We know that for any positive integer L that $\sum_{k=1}^{L} |a_k(m) - a_k(n)|^2 < \epsilon$, for all m and n larger than our earlier chosen N. By choosing m large enough, we have that

$$\sum_{k=1}^{L} |a_k - a_k(n)|^2 \le \epsilon.$$

This is true for all L, meaning that

$$\sum_{k=1}^{\infty} |a_k - a_k(n)|^2 \le \epsilon.$$

Thus the sequence of vectors (v_n) does converge to v.

We still have to show that v is in the vector space l_2. But we have

$$\sum_{k=1}^{\infty} |a_k|^2 = \sum_{k=1}^{\infty} |a_k - a_k(n) + a_k(n)|^2$$

$$\le \sum_{k=1}^{\infty} |a_k - a_k(n)|^2 + \sum_{k=1}^{\infty} |a_k(n)|^2$$

$$< \infty$$

finishing the proof.   □

Another standard complex Hilbert space is $L^2[0, 1]$, the *square integrable functions* on the interval $[0, 1]$, which is the vector space

$$L^2[0, 1] = \left\{ f : [0, 1] \to \mathbb{C} : \int_0^1 |f(x)|^2 < \infty \right\},$$

with inner product given by

$$\langle g(x), f(x)\rangle = \int_0^1 \overline{g(x)}f(x)\mathrm{d}x.$$

(This is actually not quite right, as we have to identify any two functions that are equal off of a set of measure zero; we talk about this subtlety in the last section of this chapter.)

Finally, we need to talk about bases of Hilbert spaces. A *Schauder basis* for a Hilbert space $\mathcal{H}$ is a countable collection $v_1, v_2, v_3, \ldots$ such that any vector in $\mathcal{H}$ can be written uniquely as a possibly infinite linear combination of $v_1, v_2, v_3, \ldots$. Thus, given any $v \in \mathcal{H}$, there are unique complex numbers $a_1, a_2, a_3, \ldots$ such that

$$v = \sum_{k=1}^{\infty} a_k v_k.$$

Since the preceding is an infinite series, we have to worry a little bit about what convergence means, but this is not too hard. We simply state that $v = \sum_{k=1}^{\infty} a_k v_k$ if

$$\lim_{N \to \infty} \left| v - \sum_{k=1}^{N} a_k v_k \right| = 0.$$

Since $\sum_{k=1}^{N} a_k v_k$ is a finite sum, the preceding is well-defined.

For a given Hilbert space, a basis is far from unique. But some bases are better than others, namely, those that are *orthonormal*. A basis $v_1, v_2, v_3, \ldots$ is orthonormal if

$$\langle v_i, v_j \rangle = \begin{cases} 1 & \text{if } i = j \\ 0 & \text{if } i \neq j \end{cases}.$$

For an orthonormal basis, the unique coefficients a_i have a particularly nice form:

Theorem 12.1.2. *Let* $v_1, v_2, v_3, \ldots$ *be an orthnormal basis for a Hilbert space* $\mathcal{H}$. *Then, for any vector* $v \in \mathcal{H}$, *we have*

$$v = \sum_{k=1}^{\infty} \langle v_k, v \rangle v_k.$$

Sketch of Proof: There are unique numbers a_k such that $v = \sum a_k v_k$. Then we have

$$\langle v_j, v \rangle = \langle v_j, \sum_{k=1}^{\infty} a_k v_k \rangle$$

$$= \sum_{k=1}^{\infty} \langle v_j, a_k v_k \rangle$$

$$= a_k,$$

as desired. (The reason why we called this a 'sketch of a proof' is that we did not justify why $\langle v_j, \sum_{k=1}^{\infty} a_k v_k \rangle = \sum_{k=1}^{\infty} \langle v_j, a_k v_k \rangle$. While straightforward, it does take a little work.)

Now to see that any Schauder basis can be transformed into an orthonormal basis, via the *Gram-Schmidt process*. Start with any basis $v_1, v_2, v_3, \ldots$. We will recursively construct an orthonormal basis $w_1, w_2, w_3, \ldots$. Start by setting

$$w_1 = \frac{1}{|v_1|} v_1,$$

a vector with length one.

In an intermediate step, set

$$\tilde{w}_2 = v_2 - \langle w_1, v_2 \rangle w_1.$$

As seen in the exercises, $\tilde{w}_2$ is orthogonal to w_1. To get this vector to have length one, set

$$w_2 = \frac{1}{|\tilde{w}_2|} \tilde{w}_2.$$

Now to make the inductive step. Suppose we have created an orthonormal sequence $w_1, \ldots, w_{n-1}$. Set

$$\tilde{w}_n = v_n - \sum_{k=1}^{n-1} \langle w_k, v_n \rangle w_k.$$

As shown in the exercises, the new vector $\tilde{w}_n$ has been created to be orthogonal to the vectors $w_1, \ldots w_{n-1}$. To get a vector of length one, we set

$$w_n = \frac{1}{|\tilde{w}_n|} \tilde{w}_n.$$

For our Hilbert space l_2, there is a quite simple basis, namely,

$$e_1 = (1,0,0,0,\ldots)$$
$$e_2 = (0,1,0,0,\ldots)$$
$$e_3 = (0,0,1,0,\ldots)$$
$$\vdots$$

For the Hilbert space $L^2[0,1]$, the sequence $1, e^{2\pi ix}, e^{-2\pi ix}, e^{4\pi ix}, e^{-4\pi ix}, \ldots$ is a natural choice for an orthonormal basis. Writing a function $f(x) \in L^2[1,0]$ in terms of this basis, if

$$f(x) = \sum_{-\infty}^{\infty} a_k e^{2\pi ikx},$$

then

$$a_k = \int_0^1 e^{-2\pi ikx} f(x)\, \mathrm{d}x$$

and is hence just a fancy way of talking about the Fourier series for $f(x)$. This is in part why some introductory books in quantum mechanics will hardly mention Hilbert spaces, concentrating instead on an approach based on Fourier series.

12.2. Hermitian Operators

Hilbert spaces are vector spaces. The natural type of map from a vector space to itself is a linear operator:

Definition 12.2.1. *A linear operator from a complex vector space V to itself is a map*

$$T : V \to V$$

such that for all vectors $v_1, v_2 \in V$ and complex numbers λ_1, λ_2, we have

$$T(\lambda_1 v_1 + \lambda_2 v_2) = \lambda_1 T(v_1) + \lambda_2 T(v_2).$$

We now look at linear operators on a Hilbert space.

Definition 12.2.2. *Let $\mathcal{H}$ be a Hilbert space. A linear operator $T : \mathcal{H} \to \mathcal{H}$ is* continuous *if whenever a sequence of vectors v_n in $\mathcal{H}$ converges to a vector v in $\mathcal{H}$, then $T(v_n)$ converges to $T(v)$. Thus if $\lim_{n\to\infty} v_n = v$, then*

$$\lim_{n\to\infty} T(v_n) = T(v).$$

It can be shown that all linear operators on a finite dimensional Hilbert space are continuous. This is not true for infinite dimensional Hilbert spaces. Also, often people do not say that an operator is continuous but instead say that it is *bounded*. These mean the same thing, as shown in the following theorem (which we will not prove, even though it is not that hard to show).

Theorem 12.2.1. *Let $\mathcal{H}$ be a Hilbert space. A linear operator $T : \mathcal{H} \to \mathcal{H}$ is* continuous *if and only if there is a constant B such that for all vectors $v \in \mathcal{H}$, we have*

$$|T(v)| \le B|v|.$$

Now to discuss the adjoint of a Hermitian operator. Its existence is the point of

Theorem 12.2.2. *Let $\mathcal{H}$ be a Hilbert space. Given any continuous linear operator $T : \mathcal{H} \to \mathcal{H}$, there exists an adjoint operator*

$$T^* : \mathcal{H} \to \mathcal{H}$$

such that for all $v, w \in \mathcal{H}$, we have

$$\langle w, T(v) \rangle = \langle T^*(w), v \rangle.$$

The proof that the adjoint always exists is in most texts on functional analysis, such as in chapter 12 of [58]. Further, if the linear operator is not continuous, there is still the notion of an adjoint always existing, but only after suitably restricting the domain of the operator. This is a nontrivial subtlety that we will ignore here. Thus we will always assume that an adjoint must exist. We will also see in a moment, via an example, that adjoints are hardly esoteric but in fact quite common.

We can now define Hermitian operators.

Definition 12.2.3. *A linear operator $T : \mathcal{H} \to \mathcal{H}$ is* Hermitian *if*

$$T = T^*.$$

As we will see in the next chapter, a basic assumption in quantum mechanics is that anything that can be measured must correspond to a Hermitian operator and that what can actually be measured (the number you get in a lab) must be an eigenvalue of the operator. (Technically, the numbers measured are in the spectrum of the operator. For us, we will restrict our attention to eigenvalues, leaving a discussion of the spectrum of an operator to the end of this section.) If we want to make measurements that have meaning, then these measurements should yield real numbers. While the eigenvalues for a general linear operator acting on a complex vector space could be complex numbers, the eigenvalues for a Hermitian operator must be real numbers:

Theorem 12.2.3. *The eigenvalues of a Hermitian operator are real numbers.*

Proof. Let λ be an eigenvalue with eigenvector v for a Hermitian operator $T : \mathcal{H} \to \mathcal{H}$. By definition this means $T(v) = \lambda v$. Now

$$\lambda \langle v, v \rangle = \langle v, \lambda v \rangle$$
$$= \langle v, T(v) \rangle$$
$$= \langle T^*(v), v \rangle, \quad \text{by the definition of adjoint}$$
$$= \langle T(v), v \rangle, \quad \text{since } T \text{ is Hermitian}$$
$$= \langle \lambda v, v \rangle$$
$$= \overline{\lambda} \langle v, v \rangle.$$

Since $\langle v, v \rangle \neq 0$, we must have $\lambda = \overline{\lambda}$, meaning that $\lambda \in \mathbb{R}$. $\qquad \square$

Theorem 12.2.4. *Let v be an eigenvector of a Hermitian operator T with eigenvalue λ, and let w be another eigenvector with distinct eigenvalue μ. Then v and w are orthogonal.*

Proof. We have

$$\lambda \langle w, v \rangle = \langle w, \lambda v \rangle$$
$$= \langle w, T v \rangle$$
$$= \langle T^* w, v \rangle \text{ by the definition of adjoint}$$
$$= \langle T w, v \rangle \text{ since } T \text{ is Hermitian}$$
$$= \langle \mu w, v \rangle$$
$$= \mu \langle w, v \rangle.$$

Since $\lambda \neq \mu$, we must have that $\langle w, v \rangle = 0$, as desired.

$\qquad \square$

Let us look at the particularly simple complex Hilbert space $\mathbb{C}^2$, just to see that Hermitian operators are not strange and unnatural. For vectors

$$v = \begin{pmatrix} v_1 \\ v_2 \end{pmatrix}, w = \begin{pmatrix} w_1 \\ w_2 \end{pmatrix},$$

the inner product is simply

$$< w, v > = \overline{w}^{\text{transpose}} \cdot v = (\overline{w}_1, \overline{w}_2) \begin{pmatrix} v_1 \\ v_2 \end{pmatrix} = \overline{w}_1 v_1 + \overline{w}_2 v_2.$$

An operator $T : \mathbb{C}^2 \to \mathbb{C}^2$ is given by a two-by-two matrix:

$$T = \begin{pmatrix} a & b \\ c & d \end{pmatrix}.$$

Then we have the adjoint being

$$T^* = \overline{T}^{\text{transpose}} = \begin{pmatrix} \bar{a} & \bar{c} \\ \bar{b} & \bar{d} \end{pmatrix},$$

by direct calculation. Now T will be Hermitian when $T = T^*$, or when

$$a = \bar{a}$$
$$b = \bar{c}$$
$$d = \bar{d}.$$

In particular, both elements along the diagonal, a and d, must be real numbers. Thus

$$\begin{pmatrix} 3 & 5+6i \\ 5-6i & 10 \end{pmatrix}$$

is Hermitian, while

$$\begin{pmatrix} 2 & 4+8i \\ 3+9i & 11 \end{pmatrix} \quad \text{and} \quad \begin{pmatrix} 2+3i & 6 \\ 6 & 7 \end{pmatrix}$$

are not.

Finally, we will briefly discuss the spectrum of an operator T on a Hilbert space $\mathcal{H}$.

Definition 12.2.4. *Let $T : \mathcal{H} \to \mathcal{H}$ be a linear operator. A complex number λ is in the* spectrum *of T if the operator*

$$T - \lambda I$$

is not invertible, where I is the identity map.

If λ is an eigenvalue with eigenvector v for an operator T, we have that

$$T(v) = \lambda v \Longleftrightarrow (T - \lambda I)(v) = 0.$$

Since an eigenvector cannot be the zero vector, this means that $T - \lambda I$ has a non-trivial kernel (non-trivial null-space) and hence cannot be invertible, showing

Eigenvalues of $T \subset$ Spectrum of T.

For infinite-dimensional Hilbert spaces, however, the spectrum can be strictly larger than the set of eigenvalues. In quantum mechanics, the numbers that can be obtained via making a measurement must be in the spectrum of the corresponding operator. These are nontrivial, significant subtleties that we will avoid. Most texts on functional analysis (such as [58]) treat these issues thoroughly.

12.3. The Schwartz Space

As we will see, not only are Hilbert spaces the natural spaces for quantum mechanics, but also it is often the case that we can do serious real-world calculations without ever concretely specifying what Hilbert space we are using. To some extent, this lack of specificity of the Hilbert space makes quantum mechanics into a quite flexible tool. (For a particularly clear explanation, see page 47 in section 3.2 of Folland's *Quantum Field Theory: A Tourist Guide for Mathematicians* [23].) Still, there are times when we actually need to work in a given vector space. This section deals with the Schwartz space. As a word of warning, the Schwartz space will not be a Hilbert space.

12.3.1. The Definition

Intuitively, the Schwartz space is the vector space of smooth functions from the real numbers to the complex numbers that approach zero extremely quickly. More precisely,

Definition 12.3.1. *The* Schwartz space $\mathcal{S}(\mathbb{R})$ *is the vector space of smooth functions*

$$f : \mathbb{R} \to \mathbb{C}$$

such that, for all $n, m \in \mathbb{N}$, we have

$$\lim_{x \to \pm\infty} \left| x^n \frac{d^m f}{dx^m} \right| = 0.$$

For $n = 0, m = 0$, we see that if $f \in \mathcal{S}(\mathbb{R})$ then

$$\lim_{x \to \infty} |f(x)| = \lim_{x \to -\infty} |f(x)| = 0,$$

and thus $f(x)$ does indeed approach zero for large $|x|$. But, since we also have, for $n = 1, m = 0$, that

$$\lim_{x \to \infty} |xf(x)| = \lim_{x \to -\infty} |xf(x)| = 0,$$

$|f|$ must go to zero fairly quickly in order to compensate for $|x|$ going to infinity.

We want to show that $\mathcal{S}(\mathbb{R})$ can be made into an inner product space by defining the inner product to be

$$\langle f, g \rangle = \int_{-\infty}^{\infty} \overline{f} g \, dx.$$

We will see that while it is an inner product space, $\mathcal{S}(\mathbb{R})$ is not complete and hence cannot be a Hilbert space.

But we first have to show, for all $f, g \in \mathcal{S}(\mathbb{R})$, that $\langle f, g \rangle = \int_{-\infty}^{\infty} \overline{f} g \, dx$ is finite. We first need

Lemma 12.3.1. *Let* $f \in \mathcal{S}(\mathbb{R})$. *Then*

$$\int_{-\infty}^{\infty} |f(x)|^2 \, dx < \infty.$$

This lemma is stating that functions in the Schwartz space approach zero so quickly that the area under $|f(x)|^2$ is finite. The technique behind the proof is one of the standard ways to estimate sizes of integrals.

Proof. We will show that each of the three integrals

$$\int_{-\infty}^{-1} |f(x)|^2 \, dx, \quad \int_{-1}^{1} |f(x)|^2 \, dx, \quad \int_{1}^{\infty} |f(x)|^2 \, dx$$

is finite.

Since $f(x)$ is continuous, from calculus we know that $|f(x)|^2$ is also continuous. Then the area under $|f(x)|^2$ over the bounded closed interval $[-1, 1]$ must be finite,

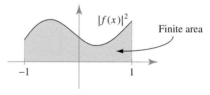

Figure 12.1

which means that

$$\int_{-1}^{1} |f(x)|^2 \, dx < \infty.$$

Now to look at $\int_{1}^{\infty} |f(x)|^2 \, dx$. We know that

$$\lim_{x \to \infty} |x f(x)| = 0.$$

This means that there is a positive constant B such that, for all $x \geq 1$, we have

$$|x f(x)| < B,$$

which in turn means that for $|x| \geq 1$,

$$|f(x)| < \left| \frac{B}{x} \right|.$$

Then

$$\int_1^\infty |f(x)|^2 \, dx < \int_1^\infty \frac{B^2}{x^2} \, dx$$
$$= B^2.$$

A similar argument, left for the exercises, will show that $\int_{-\infty}^{-1} |f(x)|^2 \, dx$ is also finite.   □

Proposition 12.3.1. *For all $f, g \in \mathcal{S}(\mathbb{R})$, we have that*

$$\langle f, g \rangle = \int_{-\infty}^\infty \overline{f} g \, dx$$

exists.

Proof. The key is that for all complex numbers z, w we have

$$|\overline{z} w| < |z|^2 + |w|^2,$$

a straightforward calculation that we leave for the exercises. Then

$$\left| \int_{-\infty}^\infty \overline{f} g \, dx \right| \leq \int_{-\infty}^\infty |\overline{f} g| \, dx$$
$$\leq \int_{-\infty}^\infty \left(|f|^2 + |g|^2 \right) \, dx$$
$$= \int_{-\infty}^\infty |f|^2 \, dx + \int_{-\infty}^\infty |g|^2 \, dx$$
$$< \infty.$$

Thus the integral $\int_{-\infty}^\infty \overline{f} g \, dx$ exists.

                                                                  □

12.3.2. The Operators $q(f) = xf$ and $p(f) = -i \, df/dx$

There are two Hermitian operators that will later be important for quantum mechanics. They are

$$q : \mathcal{S}(\mathbb{R}) \to \mathcal{S}(\mathbb{R}), \quad p : \mathcal{S}(\mathbb{R}) \to \mathcal{S}(\mathbb{R})$$

where

$$q(f)(x) = xf(x), \quad p(f)(x) = -i \frac{d}{dx}.$$

The key is

Theorem 12.3.1. *Both* $q(f)(x) = xf(x)$ *and* $p(f)(x) = -i\frac{df}{dx}$ *are Hermitian operators from* $\mathcal{S}(\mathbb{R}) \to \mathcal{S}(\mathbb{R})$.

The proof is contained in the following set of lemmas, which are left as exercises.

Lemma 12.3.2. *For any* $f \in \mathcal{S}(\mathbb{R})$, *we have that* $q(f) \in \mathcal{S}(\mathbb{R})$.

Lemma 12.3.3. *For any* $f \in \mathcal{S}(\mathbb{R})$, *we have that* $p(f) \in \mathcal{S}(\mathbb{R})$.

Lemma 12.3.4. *For all* $f, g \in \mathcal{S}(\mathbb{R})$, *we have that*

$$\langle f, q(g) \rangle = \langle q(f), g \rangle.$$

Lemma 12.3.5. *For all* $f, g \in \mathcal{S}(\mathbb{R})$, *we have that*

$$\langle f, p(g) \rangle = \langle p(f), g \rangle.$$

As we will see, the failure of the two operators to commute will be important in the quantum mechanics of springs, which in turn will be critical for the quantum mechanics of light.

Definition 12.3.2. *The* commutator *of two operators* a *and* b *is*

$$[a, b] = ab - ba.$$

Thus $[a, b] = 0$ precisely when a and b commute (meaning that $ab = ba$). To see how q and p fail to commute we need to look at $[q, p]$.

Proposition 12.3.2. *For all* $f \in \mathcal{S}(\mathbb{R})$, *we have*

$$[q, p](f)(x) = if(x).$$

Proof.

$$[q, p](f)(x) = q(p(f))(x) - p(q(f))(x)$$
$$= x\left(-i\frac{df}{dx}\right) - \left(-i\frac{d}{dx}(xf)\right)$$
$$= -ix\frac{df}{dx} - \left(-if(x) - ix\frac{df}{dx}\right)$$
$$= if(x),$$

as desired.

$\square$

12.3.3. $\mathcal{S}(\mathbb{R})$ *Is Not a Hilbert Space*

While $\mathcal{S}(\mathbb{R})$ has an inner product, it is not complete and therefore is not a Hilbert space. We will explicitly construct a Cauchy sequence of functions in $\mathcal{S}(\mathbb{R})$ that converge to a discontinuous function, and hence to a function that is not in our Schwartz space $\mathcal{S}(\mathbb{R})$.

For all positive integers n, set

$$f_n(x) = \frac{1}{e^{nx^2}} = e^{-(nx^2)}.$$

As shown in the exercises, each $f_n(x) \in \mathcal{S}(\mathbb{R})$. We will now show

Lemma 12.3.6.

$$\int_{-\infty}^{\infty} \frac{1}{e^{nx^2}}\, dx = \sqrt{\frac{\pi}{n}}.$$

Proof. This is a standard argument from calculus, using one of the cleverest tricks in mathematics. Set

$$A = \int_{-\infty}^{\infty} \frac{1}{e^{nx^2}}\, dx.$$

As it stands, it is difficult to see how to perform this integral. (In fact, there is no way of evaluating the indefinite integral

$$\int \frac{1}{e^{nx^2}}\, dx$$

in terms of elementary functions.)

Then, seemingly arbitrarily, we will find not A but instead its square:

$$A^2 = \left(\int_{-\infty}^{\infty} \frac{1}{e^{nx^2}}\, dx \right) \left(\int_{-\infty}^{\infty} \frac{1}{e^{nx^2}}\, dx \right)$$

$$= \left(\int_{-\infty}^{\infty} \frac{1}{e^{nx^2}}\, dx \right) \left(\int_{-\infty}^{\infty} \frac{1}{e^{ny^2}}\, dy \right),$$

since it does not matter what we call the symbol we are integrating over. But now the two integrals are integrating over different variables. Hence

$$A^2 = \left(\int_{-\infty}^{\infty} \frac{1}{e^{nx^2}}\, dx \right) \left(\int_{-\infty}^{\infty} \frac{1}{e^{ny^2}}\, dy \right) = \int_{-\infty}^{\infty} \int_{-\infty}^{\infty} \frac{1}{e^{n(x^2+y^2)}}\, dx\, dy.$$

We did this so that we can now use polar coordinates, setting

$$x = r\cos(\theta),\ y = r\sin(\theta),$$

giving us that

$$r^2 = x^2 + y^2,\ r\, dr\, d\theta = dx\, dy,$$

with the limits of integration being given by $0 \leq r \leq \infty$ and $0 \leq \theta \leq 2\pi$. Then we have

$$A^2 = \int_0^{2\pi} \int_0^{\infty} e^{-nr^2} r \, dr \, d\theta$$

$$= \int_0^{2\pi} \frac{-1}{2n} e^{-nr^2} \Big|_0^{\infty} \, d\theta$$

$$= \int_0^{2\pi} \frac{1}{2n} \, d\theta$$

$$= \frac{1}{2n} \int_0^{2\pi} \, d\theta$$

$$= \frac{\pi}{n},$$

finishing the proof.

$\square$

We will now show that the sequence of functions

$$f_n(x) = \frac{1}{e^{nx^2}}$$

is a Cauchy sequence in our Schwartz space that does not converge to a function in the Schwartz space. Assume for a moment that this sequence is Cauchy. Note that for a fixed number $x \neq 0$ we have

$$\lim_{n \to \infty} f_n(x) = 0.$$

For $x = 0$, though, we have for all n that

$$f_n(0) = 1.$$

Thus

$$\lim_{n \to \infty} f_n(0) = \lim_{n \to \infty} 1 = 1,$$

giving us that the limit function f of the sequence f_n is the discontinuous function

$$f(x) = \begin{cases} 1 & \text{if} \quad x = 0 \\ 0 & \text{if} \quad x \neq 0. \end{cases}$$

Hence the limit function $f(x)$ is not in $\mathcal{S}(\mathbb{R})$, meaning that $\mathcal{S}(\mathbb{R})$ cannot be a Hilbert space. Thus we must show

Lemma 12.3.7. *The sequence of functions $f_n(x) = e^{-nx^2}$ is Cauchy.*

Proof. We need to show

$$\lim_{n,m \to \infty} |f_n - f_m|^2 = 0.$$

Since the functions f_n are all real-valued, we need not worry about using conjugation signs. Then we have

$$
\begin{aligned}
|f_n - f_m|^2 &= \int_{-\infty}^{\infty} (f_n - f_m)(f_n - f_m)\, dx \\
&= \int_{-\infty}^{\infty} (f_n^2 - 2f_n f_m + f_m^2)\, dx \\
&\leq \int_{-\infty}^{\infty} f_n^2\, dx + 2\int_{-\infty}^{\infty} f_n f_m\, dx + \int_{-\infty}^{\infty} f_m^2\, dx \\
&= \int_{-\infty}^{\infty} e^{-2nx^2}\, dx + 2\int_{-\infty}^{\infty} e^{-(n+m)x^2}\, dx + \int_{-\infty}^{\infty} e^{-2mx^2}\, dx \\
&= \sqrt{\frac{\pi}{2n}} + 2\sqrt{\frac{\pi}{n+m}} + \sqrt{\frac{\pi}{2m}} \\
&\to 0,
\end{aligned}
$$

as $n,m \to \infty$, as desired. $\qquad\qquad\qquad\qquad\qquad\qquad\qquad\qquad\square$

12.4. Caveats: On Lebesgue Measure, Types of Convergence, and Different Bases

To make much of this chapter truly rigorous would require a knowledge of Lebesgue measure, a standard and beautiful topic covered in such classics as those by Royden [56], Rudin [57], and Jones [34], and by many others. For example, while our definition for the Hilbert space

$$l_2 = \left\{ (a_1, a_2, a_3, \dots) : a_k \in \mathbb{C} \text{ and } \sum_{k=1}^{\infty} |a_k|^2 < \infty \right\}$$

is rigorous, our definition for square integrable functions

$$L^2[0,1] = \left\{ f : [0,1] \to \mathbb{C} : \int_0^1 |f(x)|^2 < \infty \right\}$$

is not. While this is a vector space, it is not actually complete and hence not quite a Hilbert space. To make it into an honest Hilbert space, we have to place an equivalence relation on our set of functions. Specifically, we have to identify two functions if they are equal off of a set of measure zero (where "measure zero" is a technical term from Lebesgue measure theory).

This also leads to different types of convergence. Even in the world of continuous functions, there is a distinction between pointwise convergence and uniform convergence. With the vector space $L^2[0,1]$, we can also talk about L^2 convergence. Here we say that a sequence of functions $f_n(x) \in L^2[0,1]$ will L^2 converge to a function $f \in L^2[0,1]$ if the numbers $\int_0^1 |f_n(x) - f(x)|^2 dx$ converge to zero.

Consider our sequence of functions $f_n(x) = e^{-(nx^2)}$ in Schwartz space $\mathcal{S}(\mathbb{R})$. We showed that this sequence converged to the discontinuous function that is one at $x = 0$ and zero everywhere else. This convergence is both pointwise and in the sense of L^2. But this sequence does converge in L^2 to a perfectly reasonable function in the Schwartz space, namely, to the zero function. Note that the zero function and the function that is one at $x = 0$ and zero everywhere else, while certainly not equal pointwise, are equal off a set of measure zero. There are, though, examples of sequences in $\mathcal{S}(\mathbb{R})$ that converge pointwise and in L^2 to functions that are not in $\mathcal{S}(\mathbb{R})$.

Finally a word about bases for vector spaces. A "basis" from beginning linear algebra for a vector space V is a collection of vectors such that every vector $v \in V$ can be written uniquely as a finite linear combination of vectors from the basis. Every vector space has such a basis. This type of basis is called a *Hamel basis*. Unfortunately, for infinite-dimensional vector spaces, it can be shown that a Hamel basis has to have an uncountable number of elements. This is a deep fact, linked to the axiom of choice. It also means that such a basis is almost useless for infinite-dimensional spaces. Luckily for Hilbert spaces there is another type of basis, namely, the Schauder basis defined in this chapter.

12.5. Exercises

Exercise 12.5.1. *Find two complex numbers α and β such that*

1.
$$|\alpha + \beta|^2 > |\alpha|^2 - |\beta|^2,$$

2.
$$|\alpha + \beta|^2 = |\alpha|^2 + |\beta|^2,$$

3.
$$|\alpha + \beta|^2 < |\alpha|^2 + |\beta|^2.$$

Exercise 12.5.2. *Show for all complex numbers $z, w \in \mathbb{C}$ that*

$$|zw| \leq |z|^2 + |w|^2.$$

Exercise 12.5.3. *Let V be the vector space consisting of infinite sequences of complex numbers, with all but a finite number being zero. Show that*

$$\langle (a_n), (b_m) \rangle = \sum_{k=1}^{\infty} \overline{a}_k b_k$$

defines an inner product on V.

Exercise 12.5.4. *The goal of this exercise is to show that l_2 is a vector space.*

1. *Let $(a_k), (b_k) \in l_2$. Show that $(a_k + b_k) \in l_2$.*
2. *Let $(a_k) \in l_2$ and let $\lambda \in \mathbb{C}$. Show that $\lambda(a_k) = (\lambda a_k)$ is still in l_2.*

Exercise 12.5.5. *Let $v_1, v_2, v_3, \ldots$ be a basis for a Hilbert space $\mathcal{H}$.*

1. *Show that the vector $w_1 = \frac{1}{|v_1|} v_1$ has length one.*
2. *Show that*
   $$\tilde{w}_2 = v_2 - \langle w_1, v_2 \rangle w_1$$
   is orthogonal to w_1.
3. *Find a vector w_2 that has length one and is orthogonal to w_1, such that the vectors w_1 and w_2 span the same space as v_1 and v_2.*
4. *Suppose $w_1, \ldots, w_{n-1}$ is an orthonormal sequence. Setting*
   $$\tilde{w}_n = v_n - \sum_{k=1}^{n-1} \langle w_k, v_n \rangle w_k,$$
   show that $\tilde{w}_n$ is orthogonal to each w_k.
5. *Set*
   $$w_n = \frac{\tilde{w}_n}{|\tilde{w}_n|}.$$
   Show that $w_1, w_2, \ldots$ is an orthonormal basis for $\mathcal{H}$.

Exercise 12.5.6. *For the Hilbert space $\mathbb{C}^2$ with inner product*

$$\langle w, v \rangle = \overline{w}^{\text{transpose}} \cdot v$$

show that

$$v_1 = \begin{pmatrix} 1 \\ 2 \end{pmatrix}, v_2 = \begin{pmatrix} 3 \\ -2 \end{pmatrix}$$

form a basis. Use the Gram-Schmidt process to construct another basis that is orthonormal.

Exercise 12.5.7. *For the Hilbert space $\mathbb{C}^2$ with inner product*

$$\langle w, v \rangle = \overline{w}^{\text{transpose}} \cdot v,$$

show that the adjoint of

$$T = \begin{pmatrix} a & b \\ c & d \end{pmatrix}$$

is

$$T^* = \overline{T}^{\text{transpose}} = \begin{pmatrix} \bar{a} & \bar{c} \\ \bar{b} & \bar{d} \end{pmatrix}.$$

Exercise 12.5.8. *Consider the Hilbert space* $\mathbb{C}^3$ *with inner product* $\langle w, v \rangle = \overline{w}^{\text{transpose}} \cdot v$. *Let*

$$T = \begin{pmatrix} a_{11} & a_{12} & a_{13} \\ a_{21} & a_{22} & a_{23} \\ a_{31} & a_{32} & a_{33} \end{pmatrix}.$$

Show that T is Hermitian if and only if

$$a_{ij} = \overline{a_{ji}},$$

for all i and j.

Exercise 12.5.9. *Generalize the preceding to the vector space* $\mathbb{C}^n$.

Exercise 12.5.10. *For any* $f \in \mathcal{S}(\mathbb{R})$, *show that*

$$\int_{-\infty}^{-1} |f(x)|^2 \, dx < \infty.$$

Exercise 12.5.11. *For any* $f \in \mathcal{S}(\mathbb{R})$, *show that* $q(f) \in \mathcal{S}(\mathbb{R})$.

Exercise 12.5.12. *For any* $f \in \mathcal{S}(\mathbb{R})$, *show that* $p(f) \in \mathcal{S}(\mathbb{R})$.

Exercise 12.5.13. *For all* $f, g \in \mathcal{S}(\mathbb{R})$, *show that*

$$\langle f, q(g) \rangle = \langle q(f), g \rangle.$$

Exercise 12.5.14. *For all* $f, g \in \mathcal{S}(\mathbb{R})$, *show that*

$$\langle f, p(g) \rangle = \langle p(f), g \rangle.$$

Exercise 12.5.15. *For each positive n, show that*

$$\lim_{x \to \pm\infty} x^k e^{-(nx^2)} = 0.$$

Exercise 12.5.16. *For each positive n, show that*

$$\lim_{x \to \pm\infty} \left| x^k \frac{d^m e^{-(nx^2)}}{dx^m} \right| = 0,$$

for all m, allowing us to conclude that $e^{-(nx^2)} \in \mathcal{S}(\mathbb{R})$.

13

Some Quantum Mechanical Thinking

Summary: After discussing the photoelectric effect, which gives experimental evidence that light is not solely a wave but also must have particle properties, we will state the basic assumptions behind quantum mechanics. The last section discusses quantization, which is a method to link traditional classical physics with the new world of quantum mechanics.

13.1. The Photoelectric Effect: Light as Photons

Shining light on certain metals will cause electrons to fly off.

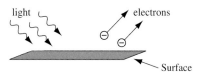

Figure 13.1

This is called the photoelectric effect. Something in light provides enough "oomph" to kick the electrons off the metal.

At a qualitative level, this makes sense. We have seen that light is an electromagnetic wave. Since waves transmit energy, surely an energetic enough light wave would be capable of kicking an electron out of its original metal home.

Unfortunately, this model quickly breaks down when viewed quantitatively. The energy of the fleeing electrons can be measured. A classical wave has energy that is proportional to the square of its amplitude (we will give an intuition for this in a moment), but changing the amplitude of the light does not change the energy of the electrons. Instead, changing the frequency of the light wave is what changes the energy of the electrons: Higher frequency means higher electron energies. This is just not how waves should behave.

In 1905, Albert Einstein made the assumption that light is not a wave but instead is made up of particles. He further assumed that the energy of each particle is $h\nu$, where h is a number called the *Planck constant* and, more importantly, ν is the frequency of the light. This assumption agreed with experiment.

Einstein made this claim for the particle nature of light the very same year that he developed his Special Theory of Relativity, which is built on the wave nature of light. This photoelectric theory paper, though, is far more radical than his special relativity paper. For an accurate historical account, see Pais's Subtle Is the Lord: The Science and the Life of Albert Einstein [51].

Even so, Einstein is still using, to some extent, a wavelike property of light, in that light's energy depends on its frequency, which is a wave property. Still, none of this agrees with the world of waves and particles in classical physics, leading by the mid-1920s to the quantum revolution.

Before giving a barebones outline for quantum mechanics in the next section, let us first look at why a wave's energy should be related to its amplitude, as opposed to its frequency.

Think of a wave as made up of a piece of string. In other words, we have a material wave. Since light is an electromagnetic wave, people thought it had to be a wave of something. This something was called the "ether."

Consider a wave moving through time, with fixed endpoints:

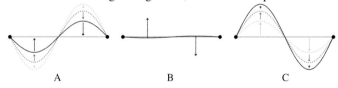

A   B   C

Figure 13.2

Now, the energy of the wave is the sum of its potential energy and its kinetic energy. The up-and-down motion of the wave, and hence the wave's kinetic energy, is at its least when the wave is stretched the most. Thus the potential energy (where the kinetic energy is zero) should be proportional to some function of the amplitude. (We will see why it is the square of the amplitude in the next chapter.) The frequency of the wave should not affect its energy, unless the wave-like nature of electromagnetism is a fundamentally different type of wave.

13.2. Some Rules for Quantum Mechanics

Historically, it took only about twenty years for people to go from Einstein's interpretation of the photoelectric effect and other experimental oddities, such as the sharp spectral lines of hydrogen, to a sharp, clean, spectacular new theory of mechanics: quantum mechanics. There are many ways to introduce the beginnings of quantum mechanics, ranging from a mechanical, rule-based scheme for calculations to a presentation based on a highly abstract axiomatics. Our underlying motivation is to identify some of the key mathematical structures behind the physics and then to take these structures seriously, even in non-physical contexts.

Thus we will present here some of the key rules behind quantum mechanics. It will not be, in any strict sense, an axiomatic presentation of quantum mechanics. (Our presentation is heavily influenced by chapter 3 in [17] and, to a lesser extent, section 2.2 in [38].)

We want to study some object. For us to be able to apply mathematical tools, we need to be able to describe this object mathematically, to measure its properties, and to predict how it will change through time. Let us look at a classical example: Suppose we want to understand a baseball of mass m flying through the air. We need to know its position and its momentum.

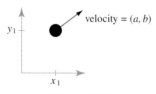

Figure 13.3

Fixing some coordinate system, we need to know six numbers: the three spatial coordinates and the three momentum coordinates. The baseball will then be described as a point

$$\left(x, y, z, m\frac{dx}{dt}, m\frac{dy}{dt}, m\frac{dz}{dt}\right) \in \mathbb{R}^6.$$

We say that the *state* of the baseball is its corresponding point in $\mathbb{R}^6$. To determine where the baseball will move through time, we use Newton's second law. We need to find all the forces F acting on the baseball and then solve the system of differential equations $F = ma$.

In quantum mechanics, we must be able to describe the state of an object and to know how to measure its properties and to find the analog of $F = ma$. None of this will be at all obvious.

The natural (though hardly naive) setting for quantum mechanics is that of Hilbert spaces and Hermitian operators. While technical definitions were given in the last chapter, you can think of a Hilbert space as a complex vector space with an inner product (and thus a vector space for which we have a notion of angle) and Hermitian operators as linear transformations on the Hilbert space.

Assumption 1. *The state of an object will be a ray in a Hilbert space* $\mathcal{H}$. *Thus a state of an object is specified by a non-zero vector* v *in the Hilbert space, and two vectors specify the same state if they are non-zero multiples of each other.*

Assumption 2. *Any quantity that can be measured will correspond to a Hermitian operator*

$$A : \mathcal{H} \to \mathcal{H}.$$

Thus, if we want to measure, say, the velocity of our object, there must correspond a "velocity" operator.

Assumption 3. *Actual measurements will always be in the spectrum of the corresponding Hermitian operator.*

A number λ will be in the spectrum of an operator A if the operator

$$A - \lambda I$$

is non-invertible, where I is the identity operator. If λ is an eigenvalue with eigenvector v for an operator A, i.e.,

$$Av = \lambda v,$$

we must have that

$$(A - \lambda I)v = 0,$$

which means that $A - \lambda I$ is non-invertible. For most practical purposes, one can think of the spectrum as the eigenvalues of A. This is why the previous assumption is frequently thought of as saying that actual measurements must be eigenvalues of the corresponding Hermitian operator. For the rest of this section, we will assume that the spectrum for each of our Hermitian operators only consists of eigenvalues.

Recall that an eigenvalue of operator A is simple if its eigenvectors form a one-dimensional subspace of the Hilbert space.

Assumption 4. *Suppose our object has state* $v \in \mathcal{H}$. *Let* A *be a Hermitian operator with simple eigenvalue* λ *and corresponding eigenvector* $w \in \mathcal{H}$. *When we make a measurement on our object for the quantity corresponding to* A, *then the probability that we get the number* λ *is*

$$\frac{|\langle w, v \rangle|^2}{\langle w, w \rangle \langle v, v \rangle}.$$

Now we start to leave the classical, Newtonian world far behind. We will take a few paragraphs to start discussing what this assumption actually entails.

We are using the Hilbert space structure, more precisely the inner product $\langle \cdot, \cdot \rangle$. First, we want to make sure that $\frac{|\langle w,v \rangle|^2}{\langle w,w \rangle \langle v,v \rangle}$ is a real number between 0 and 1 (since we want to interpret this number as a probability).

Proposition 13.2.1. *For any non-zero vector $v \in \mathcal{H}$, we have*

$$0 \leq \frac{|\langle w,v \rangle|^2}{\langle v,v \rangle \langle w,w \rangle} \leq 1.$$

Proof. Since $\mathcal{H}$ is a complex vector space, the inner product $\langle w,v \rangle$ could be a complex number, but $|\langle w,v \rangle|^2$ is a real number. Similarly, as we saw in the last chapter, while in general $\langle w,v \rangle$ does not need to be a real number, we always have that $\langle v,v \rangle$ is a positive real number (provided $v \neq 0$). Thus $\frac{|\langle w,v \rangle|^2}{\langle v,v \rangle \langle w,w \rangle}$ is a nonnegative real number.

For this number to have a chance to be a probability, we need it to be less than or equal to 1. This follows, though, from the Cauchy-Schwarz inequality, which states that

$$|\langle w,v \rangle| \leq \sqrt{\langle w,w \rangle}\sqrt{\langle v,v \rangle}.$$

$\square$

The squaring of $|\langle w,v \rangle|$ leads to non-linearity, which is built into the heart of quantum mechanics. Suppose we have two states v_1 and v_2. While

$$\langle w, v_1 + v_2 \rangle = \langle w, v_1 \rangle + \langle w, v_2 \rangle,$$

it is almost always the case that

$$|\langle w, v_1 + v_2 \rangle|^2 \neq |\langle w, v_1 \rangle|^2 + |\langle w, v_2 \rangle|^2.$$

(This is really a fact about complex number, and is shown in the exercises in chapter 12.) Though we will not be talking about it, this is the mathematical underpinning of quantum interference.

We immediately have

Proposition 13.2.2. *Suppose that our state is an eigenvector w with eigenvalue λ for a Hermitian operator A. Then the probability when we take a measurement for the value corresponding to A that we measure the number λ is precisely one.*

Finally, we defined our states to be rays in our Hilbert space. But in making measurements, we chose a vector on the ray. What if we had chosen a different vector on the ray? We want to make sure that our probabilities do not change, as the following proves:

Proposition 13.2.3. *Let A be a Hermitian operator with simple eigenvalue λ with eigenvector $w \in \mathcal{H}$. The probability that we get the value λ when we measure the state specified by v is precisely equal to the probability of getting λ when we measure the state specified by μv, for μ any complex number not equal to 0.*

Proof. The probability of measuring λ for the state μv is

$$\frac{|\langle w, \mu v \rangle|^2}{\langle w, w \rangle \langle \mu v, \mu v \rangle}.$$

We have

$$\frac{|\langle w, \mu v \rangle|^2}{\langle w, w \rangle \langle \mu v, \mu v \rangle} = \frac{|\mu \langle w, v \rangle|^2}{|\mu|^2 \langle w, w \rangle \langle v, v \rangle}$$

$$= \frac{|\langle w, v \rangle|^2}{\langle w, w \rangle \langle v, v \rangle},$$

which is just the probability that, in measuring v, we get the number λ.   □

This proposition allows us to identify the ray that describes a state with an actual vector on the ray, an identification that we will regularly make, starting in the following:

Assumption 5. *Let A be a Hermitian operator that has a simple eigenvalue λ with eigenvector $w \in \mathcal{H}$. Start with a state $v \in \mathcal{H}$. Take a measurement on the state v for the quantity corresponding to A. If we measure the number λ, then the state of our object after the measurement is now the ray spanned by the eigenvector w.*

By this assumption, the act of measurement will fundamentally change the state of the object. Further, we can only know probabilistically what the new state will be. Let us look at an example. Suppose that our Hilbert space is $\mathcal{H} = \mathbb{C}^2$ with inner product given by

$$\langle (\alpha_1, \alpha_2), (\beta_1, \beta_2) \rangle = \overline{\alpha}_1 \beta_1 + \overline{\alpha}_2 \beta_2.$$

Let A be a Hermitian operator with eigenvectors w_1 and w_2, with corresponding eigenvalues λ_1 and λ_2, and with $\lambda_1 \neq \lambda_2$. As we saw last chapter, we know that w_1 and w_2 are orthogonal and form a basis for $\mathbb{C}^2$. Further, as we have seen, we can assume that each has length 1. Let our object initially be a vector

$v \in \mathbb{C}^2$. Then there exist complex numbers α_1 and α_2 such that

$$v = \alpha_1 w_1 + \alpha_2 w_2.$$

Suppose we measure the number λ_1. This will happen with probability

$$\frac{|\alpha_1|^2}{|\alpha_1|^2 + |\alpha_2|^2},$$

since

$$\frac{|\langle w_1, v\rangle|^2}{\langle v, v\rangle} = \frac{|\langle w_1, \alpha_1 w_1 + \alpha_2 w_2\rangle|^2}{\langle \alpha_1 w_1 + \alpha_2 w_2, \alpha_1 w_1 + \alpha_2 w_2\rangle}$$

$$= \frac{|\alpha_1|^2}{|\alpha_1|^2 + |\alpha_2|^2}.$$

Note that we are not thinking that the object is initially in either the state w_1 or the state w_2, or that the probabilities are just measuring our uncertainties. No, the assumption is that the initial state is $v = \alpha_1 w_1 + \alpha_2 w_2$. The act of measurement transforms the initial state into the state w_1 with probability $\frac{|\alpha_1|^2}{|\alpha_1|^2 + |\alpha_2|^2}$ or into the state w_2 with probability $\frac{|\alpha_2|^2}{|\alpha_1|^2 + |\alpha_2|^2}$. Chance lies at the heart of measurement.

This assumption also means that the order in which we take measurements matters. Suppose we have two Hermitian operators A and B. If we take a measurement corresponding to A, the state of the system must now become an eigenvector of A, while if we take a measurement corresponding to B, the state of the system must be an eigenvector of B. If we first measure A and then B, we will end up with the state being an eigenvector of B, while if we first measure B and then A, our state will end up being an eigenvector of A. Thus the order of measuring matters, since the eigenvectors of A and B may be quite different. This is far from the classical world.

The previous assumptions are all about how to calculate measurements in quantum mechanics. The next assumption is about how the state of an object can evolve over time and is hence the quantum mechanical analog of Newton's second law $F = ma$.

Assumption 6. *Let $v(t) \in \mathcal{H}$ be a function of time t describing the evolution of an object. There is a Hermitian operator $H(t)$ such that $v(t)$ must satisfy the partial differential equation*

$$i\hbar \frac{\partial v(t)}{\partial t} = H(t)v(t).$$

This partial differential equation is called Schrödinger's equation.

The constant $\hbar$ is Planck's constant divided by 2π. More importantly, the Hermitian operator $H(t)$ is called the *Hamiltonian* and corresponds to the classical energy. Hence its eigenvalues will give the energy of the object.

At this level of abstraction, it is hard to see how one could ever hope to set up an actual partial differential equation (PDE) that could be solved, so let us look at a specific example. (Though standard, we are following section 5.1 of [55].) Suppose we have a particle restricted to moving on a straight line, with coordinate x. The Hilbert space will be a suitable space of complex-valued functions on $\mathbb{R}$. Hence our states will be functions $v(x)$. Since our states will be evolving in time, we will actually have two-variable functions $v(x,t)$. We assume that our object has a mass m and let $U(x)$ be some function corresponding to the potential energy. Then the Hamiltonian will be

$$H(x,t)v(x,t) = - \left(\frac{\hbar}{2m} \right) \frac{\partial^2 v}{\partial x^2} + U(x)v(x,t),$$

giving us Schrödinger's equation:

$$i\hbar \frac{\partial v(x,t)}{\partial t} = - \left(\frac{\hbar}{2m} \right) \frac{\partial^2 v}{\partial x^2} + U(x)v(x,t).$$

The reader should not feel that the preceding is reasonable; we stated it just to show that real physics can turn Assumption 6 into a concrete equation we can sometimes solve.

Also, we are being deliberately cavalier about the nature of our Hilbert space of functions. Frequently we will look at the Hilbert space of square-integrable functions with respect to Lebesgue measure. These functions need not be continuous, much less differentiable. Our goal in this section was not to deal with these (interesting) issues but instead to give an outline of the underlying mathematical structure for quantum mechanics.

13.3. Quantization

Classical Newtonian mechanics accurately describes much, but not all, of the world around us. *Quantization* is the procedure that allows us to start from a Newtonian description of a system and produce a quantum mechanical one. We will explicitly do this in the next chapter with harmonic oscillators (springs).

In classical physics, we need coordinate systems. If we have n particles, then each particle will be described by its position and momentum. Positions and momentums can be measured in the lab. To move to quantum mechanics, we need to replace the classical positions and momentums with operators. But which operators do we chose? This process of replacing classical variables with appropriate operators is the goal of quantization. The key is that in general operators do not commute. We must choose our operators so that their commutators have certain specified values. (Recall that given operators a and b, the *commutator* is $[a,b] = ab - ba$.) Using deep analogies with classical mechanics (in particular with something called the Poisson bracket), Dirac, in chapter IV of his classic *Principles of Quantum Mechanics* [18], developed the following method.

Classically, for n particles, the positions describe a point in $\mathbb{R}^{3n}$ (since each particle will have x, y, and z coordinates), and the corresponding momentums describe a different point in $\mathbb{R}^{3n}$ (since the momentum of each particle will also have x, y, and z coordinates). Thus a system of n particles is described classically by a point in $\mathbb{R}^{6n}$.

To quantize, we replace each of the position coordinates by operators $q_1,\ldots,q_{3n}$ and each of the momentum coordinates by operators $p_1,\ldots,p_{3n}$, subject to the commutator rules

$$[q_i, q_j] = 0$$

$$[p_i, p_j] = 0$$

and

$$[q_i, p_j] = \begin{cases} i\hbar & \text{if } i = j \\ 0 & \text{if } i \neq j \end{cases}.$$

For Dirac, each of these commutator rules stem from corresponding classical properties of the Poisson bracket. (We are not defining the Poisson bracket but instead are mentioning them only to let the reader know that these commutator rules were not arrived at arbitrarily.) Of course, we have not come close to specifying the underlying Hilbert space. Frequently, quite a lot of information can be gleaned by treating the entire system at this level of abstraction.

Let us make all of this a bit more specific. Consider a one-dimensional system, where the Hilbert space is some suitable space of differentiable functions on the interval $[0,1]$. We will have one coordinate operator q and one momentum operator p. Letting $f(x)$ be a function in our Hilbert space, we define

$$q(f) := xf(x),$$

or, in other words, simply multiplication by x. In order for $[q, p] = i\hbar$ to hold, we need

$$p(f) := -i\hbar \frac{\mathrm{d}f}{\mathrm{d}x},$$

since

$$\left[x, -i\hbar \frac{\mathrm{d}}{\mathrm{d}x} \right] = i\hbar f.$$

It is not uncommon for the position operators to be multiplications by the traditional coordinates; that, in turn, forces the momentums to be $-i\hbar$ times the corresponding partial derivative.

13.4. Warnings of Subtleties

A rigorous axiomatic development of quantum mechanics is non-trivial. This is why we said we were making "assumptions," as opposed to stating axioms. Our assumptions are closer to "rules of thumb." For example, frequently the Hermitian operator corresponding to an observable will not be defined on the entire Hilbert space but instead on a dense subspace. Often, the operators will be defined on a Schwartz space, which, as we saw in the last chapter, is not itself a Hilbert space.

13.5. Exercises

Exercise 13.5.1. *Consider the Hilbert space* $\mathbb{C}^2$ *with inner product*

$$\langle w, v \rangle = \overline{w}^{\text{transpose}} \cdot v.$$

Show that

$$A = \begin{pmatrix} 2 & 0 \\ 0 & 3 \end{pmatrix}$$

is a Hermitian operator. Let

$$v = \begin{pmatrix} 1 \\ 2 \end{pmatrix}$$

be a vector corresponding to the state of some object. Suppose we take a measurement of v *for the quantity corresponding to* A.

1. *What is the probability that we get a measurement of 2?*
2. *What is the probability that we get a measurement of 3?*
3. *What is the probability that we get a measurement of 4?*

Exercise 13.5.2. *Consider the Hilbert space* $\mathbb{C}^2$ *with inner product*

$$\langle w, v \rangle = \overline{w}^{\text{transpose}} \cdot v.$$

Show that

$$A = \begin{pmatrix} 3 & 2+i \\ 2-i & 1 \end{pmatrix}$$

is a Hermitian operator. Find the eigenvectors and eigenvalues of A. Let

$$v = \begin{pmatrix} 1 \\ 2 \end{pmatrix}$$

be a vector corresponding to the state of some object. Suppose we make a measurement of v for the quantity corresponding to A.

1. *What is the probability that we get a measurement of $2 + \sqrt{6}$?*
2. *What is the probability that we get a measurement of 3?*

Exercise 13.5.3. *1. Let*

$$A = \begin{pmatrix} a_{11} & a_{12} \\ a_{21} & a_{22} \end{pmatrix}.$$

Show that

$$P(x) = \det(A - xI)$$

is at most a degree two polynomial.
2. *Let*

$$A = \begin{pmatrix} a_{11} & a_{12} & a_{13} \\ a_{21} & a_{22} & a_{23} \\ a_{31} & a_{32} & a_{33} \end{pmatrix}.$$

Show that

$$P(x) = \det(A - xI)$$

is at most a degree three polynomial.
3. *Let A be an $n \times n$ matrix. Show that*

$$P(x) = \det(A - xI)$$

is at most a degree n polynomial.

Exercise 13.5.4. *Let A be an $n \times n$ matrix. Let λ be an eigenvalue of A. Show that λ must be a root of*

$$P(x) = \det(A - \lambda I).$$

Conclude that A can have at most n eigenvalues.

Exercise 13.5.5. *Let*

$$A = \begin{pmatrix} 3 & 2 \\ 2 & 5 \end{pmatrix}.$$

Find the eigenvectors and eigenvalues of A.

Exercise 13.5.6. *Let*

$$B = \begin{pmatrix} 1 & 0 \\ 0 & 2 \end{pmatrix}.$$

1. *Find the eigenvalues and eigenvectors of B.*
2. *Using the matrix A in the previous problem, show that*

$$AB \neq BA.$$

3. *Show that AB has different eigenvectors than BA.*
4. *Explain why measuring with respect to B first, then with respect to A is fundamentally different from measuring with respect to A first, then with respect to B. Hence the order in which we take measurements matters.*

Exercise 13.5.7. *Let A and B be n × n matrices such that A has n distinct eigenvalues. Suppose that*

$$AB = BA.$$

Show that B has the same eigenvectors.

Exercise 13.5.8. *Let $\mathcal{H}$ be a Hilbert space. Let A be a Hermitian operator acting on $\mathcal{H}$ and let $v \in \mathcal{H}$. Let μ be any nonzero complex number. Let λ be an eigenvalue of A with eigenvector w. Show that the probability of measuring the number λ for the state corresponding to v is the same probability as measuring the number λ for the state corresponding to μv. Explain why there is no way of distinguishing the state v from the state μv.*

Exercise 13.5.9. *Let V be a vector space of infinitely differentiable functions on the unit interval. Let $f(x) \in V$.*

1. *Show that the operator*

$$q(f) = xf(x)$$

   is a linear operator.
2. *Show that the operator*

$$p(f) = -i\hbar \frac{\mathrm{d}(f)}{\mathrm{d}x}$$

   is a linear operator.
3. *Show that*

$$[q,q] = 0$$
$$[p,p] = 0$$
$$[q,p] = i\hbar.$$

Exercise 13.5.10. *Let V be a vector space of infinitely differentiable functions on the unit square. Let $f(x_1, x_2) \in V$.*

1. *Show that the operators*

$$q_1(f) = x_1 f(x_1, x_2)$$

and

$$q_2(f) = x_2 f(x_1, x_2)$$

are linear operators.
2. *Show that the operators*

$$p_1(f) = -i\hbar \frac{\partial(f)}{\partial x_1}$$

and

$$p_2(f) = -i\hbar \frac{\partial(f)}{\partial x_2}$$

are linear operators.
3. *Show that*

$$[q_i, q_j] = 0$$
$$\left[p_i, p_j\right] = 0$$
$$\left[q_i, p_i\right] = i\hbar$$
$$\left[q_i, p_j\right] = 0 \text{ for } i \neq j.$$

Exercise 13.5.11. *Let V be the set of all continuous functions*

$$f : [0, 1] \to \mathbb{C}.$$

(You may assume that V is a complex vector space.) Define

$$\langle \cdot, \cdot \rangle : V \times V \to \mathbb{C}$$

by setting

$$\langle f, g \rangle = \int_0^1 \overline{f(x)} g(x)\, dx.$$

Show that this is an inner product.

Exercise 13.5.12. *Show that the preceding vector space V is not complete.*

14

Quantum Mechanics of Harmonic Oscillators

Summary: The goal for this chapter is to quantize harmonic oscillators (e.g., springs). The underlying mathematics of harmonic oscillators is key to understanding light, as we will see in the next chapter. We will see that the possible energies for a quantized harmonic oscillator can only occur in discrete (quantized) values.

14.1. The Classical Harmonic Oscillator

It is surprising how much of the world can be modeled by using harmonic oscillators. Thus we want to understand the movement of springs and pendulums, each moving without friction.

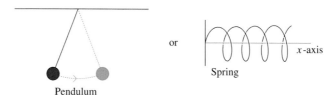

or

Pendulum

Spring

Figure 14.1

We will first give the Newtonian second law approach and then give the Lagrangian and Hamiltonian approach for understanding how a spring will move through time, allowing us to proceed to quantization in the next section.

Consider a spring

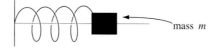

mass m

Figure 14.2

with a mass m attached at the end. The key assumption (Hooke's law), one that needs to be verified experimentally, is that the force acting on the end of the spring is

$$\text{Force} = -kx,$$

where k is a positive constant depending on the physical make-up of the spring and x is the position of the mass. Newton's second law can now be applied. Let $x(t)$ denote the position of the mass at time t. By Newton's second law, we know that $x(t)$ must satisfy the differential equation

$$m\frac{d^2 x}{dt^2} = -kx(t),$$

or

$$m\frac{d^2 x}{dt^2} + kx(t) = 0.$$

As this is a simple second order ordinary differential equation with constant coefficients, a basis for its solutions can be easily determined. One basis of solutions is

$$\cos\left(\sqrt{\frac{k}{m}}\, t\right), \sin\left(\sqrt{\frac{k}{m}}\, t\right)$$

while another is

$$\exp\left(i\sqrt{\frac{k}{m}}\, t\right), \exp\left(-i\sqrt{\frac{k}{m}}\, t\right).$$

Note that a solution of the form

$$x(t) = c_1 \cos\left(\sqrt{\frac{k}{m}}\, t\right) + c_2 \sin\left(\sqrt{\frac{k}{m}}\, t\right)$$

for constants c_1 and c_2, will indeed oscillate back and forth, just as one would expect for a frictionless spring.

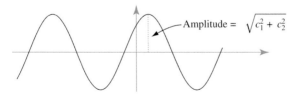

Amplitude $= \sqrt{c_1^2 + c_2^2}$

Figure 14.3

Also, using that $\sin(a+b) = \sin(a)\cos(b) + \sin(b)\cos(a)$, any solution can be written in the form

$$x(t) = A\sin\left(\sqrt{\frac{k}{m}}\, t + \phi\right),$$

for a constant ϕ. Since sine can only have values between -1 and 1, the largest that $x(t)$ can be is the absolute value of A, which is the *amplitude*.

Now to find the Hamiltonian, which is

$$\text{Hamiltonian} = H = \text{Kinetic Energy} + \text{Potential Energy}.$$

The kinetic energy is the term that captures the energy involving speed. Technically,

$$\text{Kinetic Energy} = \frac{1}{2}mv^2 = \frac{1}{2m}p^2,$$

where v is the velocity and $p = mv$ is the momentum.

The potential energy is the term that gives us the energy depending on position. Technically, its definition is

$$\text{Potential Energy} = -\int_0^x \text{Force } dx,$$

assuming that we start with zero potential energy. Since for a harmonic oscillator the force is $-kx$, we have

$$\text{Potential Energy} = -\int_0^x -kt \; dt = \frac{k}{2}x^2.$$

Thus we have that the total energy is

$$H = \frac{1}{2m}p^2 + \frac{k}{2}x^2.$$

Now we can finally show that the energy is proportional to the square of the amplitude.

Theorem 14.1.1. *The total energy for a harmonic oscillator with solution* $x(t) = A\sin\left(\sqrt{\frac{k}{m}}\,t + \phi\right)$ *is* $(1/2)kA^2$.

Proof. We have

$$
\begin{aligned}
H &= \frac{1}{2}mv^2 + \frac{k}{2}x^2 \\
&= \frac{1}{2}m\left(\frac{\mathrm{d}x}{\mathrm{d}t}\right)^2 + \frac{k}{2}x^2 \\
&= \frac{1}{2}m\left(-A\sqrt{\frac{k}{m}}\cos\left(\sqrt{\frac{k}{m}}\,t + \phi\right)\right)^2 + \frac{k}{2}\left(A\sin\left(\sqrt{\frac{k}{m}}\,t + \phi\right)\right)^2 \\
&= \frac{kA^2}{2}\left(\cos^2\left(\sqrt{\frac{k}{m}}\,t + \phi\right) + \sin^2\left(\sqrt{\frac{k}{m}}\,t + \phi\right)\right) \\
&= \frac{kA^2}{2},
\end{aligned}
$$

as desired.

□

Finally, as seen in the earlier chapter on Lagrangians, we have

$$
\text{Lagrangian} = \text{Kinetic Energy} - \text{Potential Energy}
$$
$$
= \frac{1}{2m}p^2 - \frac{k}{2}x^2.
$$

14.2. The Quantum Harmonic Oscillator

(While this section will be covering standard material, we will be closely following the outline and the notation of chapter 2 in Milonni's *The Quantum Vacuum: An Introduction to Quantum Electrodynamics* [41].) From last section, we have our classical description for a harmonic oscillator's Hamiltonian:

$$
H = \frac{1}{2m}p^2 + \frac{k}{2}x^2,
$$

where m is the mass, k a positive constant, x the position, and p the momentum. We want to quantize this. In particular, we want to find the allowable energies for a quantum spring, which means we should replace in the Hamiltonian H the position variable x and the momentum variable p by operators and then find this new H's eigenvalues. As a side benefit, the all-important annihilation and creation operators will be defined, and we will begin to see the strange quantum properties of the vacuum.

Before finding these operators, we need to specify the Hilbert space, which will be an infinite-dimensional vector space of one-variable functions that converge extremely rapidly to zero for $x \to \pm\infty$. Slightly more specifically, our Hilbert space will contain

$$\mathcal{S}(\mathbb{R}) = \left\{ f : \mathbb{R} \to \mathbb{C} : \forall n, m \in \mathbb{N}, \ \lim_{x \to \pm\infty} \left| x^n \frac{d^m f}{dx^m} \right| = 0, \ \int_{-\infty}^{\infty} |f|^2 < \infty \right\},$$

the Schwartz space from the chapter on Hilbert spaces. We are deliberately being a bit vague in stating what type of functions $f : \mathbb{R} \to \mathbb{C}$ we are considering, as this is technically difficult. The inner product is

$$\langle f, g \rangle = \int_{-\infty}^{\infty} \overline{f} g \, dx.$$

The x will become the operator q that sends each function $f(x)$ to the new function $xf(x)$. Thus

$$q(f) = xf(x).$$

As seen in the Hilbert space chapter, this q is a linear operator. The operator corresponding to the momentum p, which we will also denote by p, must satisfy the commutation relation

$$[q, p] = i\hbar.$$

Also, as seen in the Hilbert space chapter, this means that as an operator

$$p = \frac{\hbar}{i} \frac{d}{dx} = -i\hbar \frac{d}{dx}.$$

We know from Assumption 3 of the previous chapter that the allowable energies will be the eigenvalues of the operator

$$H = \frac{1}{2m} p^2 + \frac{k}{2} q^2.$$

For notational convenience, which will be clearer in a moment, we set

$$\omega = \sqrt{\frac{k}{m}},$$

so that $k = m\omega^2$. Note that in the classical world, ω is the frequency of the solution.

Thus we want to understand

$$H = \frac{1}{2m} p^2 + \frac{m\omega^2}{2} q^2.$$

We are going to show that the eigenvalues and eigenvectors are indexed by the nonnegative integers $n = 0, 1, 2, \ldots$ with eigenvalues

$$E_n = \left(n + \frac{1}{2} \right) \hbar\omega.$$

Further, in showing this, we will need two associated operators: the creation and annihilation operators (also called the raising and lowering operators), which have independent and important meaning.

To start our calculations, set

$$a = \frac{1}{\sqrt{2m\hbar\omega}}(p - im\omega q).$$

We call this operator, for reasons that will be made clear in a moment, the annihilation operator or the lowering operator. Its adjoint is

$$a^* = \frac{1}{\sqrt{2m\hbar\omega}}(p + im\omega q),$$

which is to be shown in the exercises and is called the creation or raising operator. By direct calculation (also to be shown in the exercises), we have

$$[a, a^*] = 1.$$

This allows us to rewrite our Hamiltonian operator as

$$H = \frac{1}{2}\hbar\omega(aa^* + a^*a)$$

$$= \hbar\omega\left(a^*a + \frac{1}{2}\right).$$

We will concentrate on the operator a^*a. Denote an eigenvector of a^*a by v_λ, normalized to have length one ($\langle v_\lambda, v_\lambda \rangle = 1$), with eigenvalue λ. We will show in the next few paragraphs that these eigenvalues λ are nonnegative integers. We first show that these eigenvalues are non-negative real numbers.

Lemma 14.2.1. *The eigenvalues of a^*a are non-negative real numbers.*

Proof. By the definition of the adjoint a^* and of v_λ and λ, we have

$$\langle av_\lambda, av_\lambda \rangle = \langle v_\lambda, a^*av_\lambda \rangle = \lambda\langle v_\lambda, v_\lambda \rangle = \lambda.$$

Since this is the square of the length of the vector av_λ, we see that λ must be a nonnegative real number. □

Lemma 14.2.2. *Assuming that each eigenvalue of a^*a has only a one-dimensional eigenspace, then*

$$av_\lambda = \sqrt{\lambda}v_{\lambda-1}$$

and

$$a^*v_\lambda = \sqrt{\lambda+1}v_{\lambda+1}.$$

(This is why some call a a lowering operator and a^* a raising operator.)

Proof. As already mentioned, we know that

$$1 = [a, a^*] = aa^* - a^*a.$$

Then

$$a^*a = aa^* - 1.$$

Hence

$$(a^*a)av_\lambda = (aa^* - 1)av_\lambda$$
$$= aa^*av_\lambda - av_\lambda$$
$$= (\lambda - 1)av_\lambda.$$

Thus av_λ is an eigenvector of a^*a with eigenvalue $\lambda - 1$. By our assumption that each eigenvalue of a^*a has only a one-dimensional eigenspace, we have that av_λ must be a multiple of $v_{\lambda-1}$. That is,

$$av_\lambda = Cv_{\lambda-1}.$$

We know, however, that the square of the length of av_λ is λ. Since $v_{\lambda-1}$ has length one, we see that

$$C = \sqrt{\lambda}.$$

$\square$

(At the end of this section, we will show that each eigenvalue of a^*a actually does have a one-dimensional eigenspace.)

Proposition 14.2.1. *The eigenvalues λ for the operator a^*a are non-negative integers.*

Proof. By induction we have

$$a^k v_\lambda = \sqrt{\lambda}\sqrt{\lambda - 1} \cdots \sqrt{\lambda - (k+1)} v_{\lambda-k}.$$

But all of the eigenvalues are non-negative. This forces λ to be a non-negative integer.

$\square$

This allows us to write the eigenvectors as v_n, for non-negative integers n.

We are not directly interested in the eigenvectors and eigenvalues of the "auxilary" operator a^*a but instead want to know the possible energies (the eigenvalues of the Hamiltonian) of a harmonic oscillator. Hence the following is key:

Theorem 14.2.1. *The eigenvalues for the Hamiltonian for the quantum harmonic oscillator are*

$$E_n = (n + 1/2)\hbar\omega$$

with corresponding eigenvectors v_n.

Proof. We have that

$$H v_n = \hbar\omega \left(a^* a + \frac{1}{2} \right) v_n$$

$$= (n + 1/2)\hbar\omega v_n.$$

Thus the v_n are indeed eigenvectors with the desired eigenvalues.

Now to show that these are the only eigenvectors. Let w be an eigenvector of H, with corresponding eigenvalue λ. We have

$$a^* a = \frac{1}{\hbar\omega} H - \frac{1}{2}.$$

Then

$$a^* a w = \left(\frac{1}{\hbar\omega} H - \frac{1}{2} \right) w$$

$$= \left(\frac{\lambda}{\hbar\omega} - \frac{1}{2} \right) w.$$

Thus w is also an eigenvector for $a^* a$ and hence must be one of the v_n. Thus the eigenvectors for H are precisely the vectors v_n. $\qquad\square$

This result is remarkable. The energies can only be the numbers

$$(1/2)\hbar\omega, (3/2)\hbar\omega, (5/2)\hbar\omega, \ldots.$$

Further, there is a lowest energy $(1/2)\hbar\omega$. It is reasonable to think of this as the energy when nothing is happening in the system. In the quantum world, a spring that is just sitting there must have a positive energy. We will see in the next chapter that this leads in the quantum world to the vacuum itself being complicated and problematic.

We have one final technicality to deal with, namely, showing that the eigenspaces of $a^* a$ are all one-dimensional.

Lemma 14.2.3. *The eigenspaces for the operator $a^* a$ have dimension one.*

Proof. There are parts of the preceding argument that do not depend on the eigenspaces of $a^* a$ having dimension one. We used the notation that the v_λ are eigenvectors of $a^* a$ with eigenvalue λ, without requiring λ to be an integer.

We showed both that av_λ is an eigenvector of a^*a with eigenvalue $\lambda - 1$ and that each av_λ has length $\sqrt{\lambda}$. This is true for any eigenvector of length one with eigenvalue λ. Then av_λ must be $\sqrt{\lambda}$ times an eigenvector of length one with eigenvalue $\lambda - 1$. Hence $a^k v_\lambda$ must be $\sqrt{\lambda}\sqrt{\lambda - 1}\cdots\sqrt{\lambda - (k+1)}$ times an eigenvector of length one with eigenvalue $\lambda - k$. But these can never be negative. Hence each λ is a non-negative integer. In particular, 0 is an eigenvalue.

We will show that the eigenspace for the eigenvalue 0 is one-dimensional. Since the a^* and a operators map eigenspaces to each other, this will give us our result.

Here we need actually to look at the relevant Hilbert space of square-integrable one-variable functions, with some fixed interval as domain. Let v_0 be an eigenvector of a^*a with eigenvalue 0.

Then $av_0 = 0$. We will now write v_0 as a function $\psi(x)$ in our Hilbert space. Using that

$$a = p - im\omega x = \frac{\hbar}{i}\frac{d}{dx} - im\omega x,$$

we have $\psi(x)$ satisfying the first-order differential equation

$$\frac{\hbar}{i}\frac{d\psi(x)}{dx} = im\omega x\,\psi(v).$$

But under the normalization that $\psi(x)$ has length one, this first-order ordinary differential equation has the unique solution

$$\psi(x) = \left(\frac{m\omega}{4\pi\hbar}\right)^{1/4} e^{-\frac{m\omega x^2}{2\hbar}},$$

giving us our one-dimensional eigenspace.

□

14.3. Exercises

Exercise 14.3.1. *Show that*

$$x(t) = c_1 \cos\left(\sqrt{\frac{k}{m}}\,t\right) + c_2 \sin\left(\sqrt{\frac{k}{m}}\,t\right),$$

for any constants c_1 and c_2, is a solution to

$$m\frac{d^2 x}{dt^2} + kx(t) = 0.$$

Then show that

$$x(t) = c_1 \exp\left(i\sqrt{\frac{k}{m}}\,t\right) + c_2 \exp\left(-i\sqrt{\frac{k}{m}}\,t\right),$$

for any constants c_1 and c_2, are also solutions.

Exercise 14.3.2. *Let a be the annihilation operator and a^* the creation operator for the quantum harmonic oscillator. Show that*

$$[a, a^*] = 1.$$

Exercise 14.3.3. *Let a be the annihilation operator, a^* the creation operator, and H the Hamiltonian operator for the quantum harmonic oscillator. Show that*

$$H = \frac{1}{2}\hbar\omega(aa^* + a^*a)$$

$$= \hbar\omega\left(a^*a + \frac{1}{2}\right).$$

Exercise 14.3.4. *On the Schwartz space, show that*

$$\langle f, g \rangle = \int_{-\infty}^{\infty} \overline{f} g \, dx$$

defines an inner product.

Exercise 14.3.5. *Show that $q(f(x)) = xf(x)$ defines a map*

$$q : \mathcal{S}(\mathbb{R}) \to \mathcal{S}(\mathbb{R}).$$

Exercise 14.3.6. *Show that $q(f(x)) = xf(x)$ is Hermitian.*

Exercise 14.3.7. *Show that $p(f) = -i\hbar\frac{df}{dx}$ defines a map*

$$p : \mathcal{S}(\mathbb{R}) \to \mathcal{S}(\mathbb{R}).$$

Exercise 14.3.8. *Show that $p(f) = -i\hbar\frac{df}{dx}$ is Hermitian.*

Exercise 14.3.9. *Show that the creation operator $\frac{1}{\sqrt{2m\hbar\omega}}(p + im\omega q)$ is the adjoint of the annihilation operator a, using the Hilbert space of Section 2.*

Exercise 14.3.10. *Let $\mathcal{H}$ be a Hilbert space with inner product $\langle \cdot, \cdot \rangle$. Let $A, B : \mathcal{H} \to \mathcal{H}$ be Hermitian. Show that A^2 and $A + B$ are also Hermitian.*

Exercise 14.3.11. *Prove that*

$$H = \frac{1}{2m}p^2 + \frac{m\omega^2}{2}q^2$$

is Hermitian.

15

Quantizing Maxwell's Equations

Summary: Our goal is to quantize Maxwell's equations, leading to a natural interpretation of light as being composed of photons. The key is that the quantization of harmonic oscillators (or, more prosaically, the mathematics of springs) is critical to understanding light. In particular, we show that the possible energies of light form a discrete set, linked to the classical frequency, giving us an interpretation for the photoelectric effect.

15.1. Our Approach

From Einstein's explanation of the photoelectric effect, light seems to be made up of photons, which in turn are somewhat particle-like. Can we start with the classical description of light as an electromagnetic wave solution to Maxwell's equations when there are no charges or currents, quantize, and then get something that can be identified with photons? That is the goal of this chapter. (As in the previous chapter, while all of this is standard, we will be following the outline given in chapter 2 of [41].)

Maxwell's equations are linear. We will concentrate on the 'monochromatic' solutions, those with fixed frequency and direction. We will see that the Hamiltonians of these monochromatic solutions will have the same mathematical structure as a Hamiltonian for a harmonic oscillator. We know from the last chapter how to quantize the harmonic oscillator. More importantly, we saw that there are discrete eigenvalues and eigenvectors for the corresponding Hamiltonian operator. When we quantize the Hamiltonians for the monochromatic waves in an analogous fashion, we will again have discrete eigenvalues and eigenvectors. It is these eigenvectors that we can interpret as photons.

15.2. The Coulomb Gauge

We know that light is an electromagnetic wave satisfying Maxwell's equations when there are no charges ($\rho = 0$) and zero current ($j = \langle 0, 0, 0 \rangle$):

$$\nabla \cdot E = 0$$

$$\nabla \times E = -\frac{\partial B}{\partial t}$$

$$\nabla \cdot B = 0$$

$$\nabla \times B = \frac{\partial E}{\partial t},$$

where we assume that the speed of light is one. In Chapter 7, we saw that there is a vector potential field $A(x, y, z, t)$ with

$$B = \nabla \times A$$

and a scalar potential function $\phi(x, y, z, t)$ with

$$\nabla \cdot \phi = -E - \frac{\partial A}{\partial t}.$$

As discussed in Chapter 7, neither potential is unique. In order to make the problem of quantizing electromagnetic waves analogous to quantizing harmonic oscillators, we need to choose particular potentials. We will assume that "at infinity" both potentials are zero.

Theorem 15.2.1. *For an electric field E and a magnetic field B when there are no charges and current, we can choose the vector potential A to satisfy*

$$\nabla \cdot A = 0$$

(such an A is called a Coulomb potential, or a Coulomb gauge) and a scalar potential to satisfy

$$\phi = 0.$$

We will show that the part analogous to the Hamiltonian of a harmonic oscillator will be in terms of this potential.

Proof. From Chapter 7, we know that there is at least one vector potential field A such that

$$B = \nabla \times A.$$

Fix one of these solutions. Also, from Chapter 7, we know that $A + \nabla(f)$ is also a vector potential, where $f(x, y, z, t)$ is any given function.

To find a Coulomb gauge, we must show that there exists a function f such that

$$\nabla \cdot (A + \nabla(f)) = 0.$$

Thus we must find a function f such that

$$\nabla \cdot \nabla(f) = -\nabla \cdot A.$$

Set $A = (A_1, A_2, A_3)$. Since

$$\nabla \cdot \nabla(f) = \left(\frac{\partial}{\partial x}, \frac{\partial}{\partial y}, \frac{\partial}{\partial z} \right) \cdot \left(\frac{\partial f}{\partial x}, \frac{\partial f}{\partial y}, \frac{\partial f}{\partial z} \right) = \frac{\partial^2 f}{\partial x^2} + \frac{\partial^2 f}{\partial y^2} + \frac{\partial^2 f}{\partial z^2},$$

we must solve

$$\frac{\partial^2 f}{\partial x^2} + \frac{\partial^2 f}{\partial y^2} + \frac{\partial^2 f}{\partial z^2} = -\frac{\partial A_1}{\partial x} - \frac{\partial A_2}{\partial y} - \frac{\partial A_3}{\partial z},$$

where the right hand side is given. This partial differential equation can always be solved. (For background, see chapter 2, section C of [23].)

Thus we can assume that there is a vector potential A with both $B = \nabla \times A$ and $\nabla \cdot A = 0$. We must now show that we can choose the scalar potential ϕ also to be identically zero. Here we will need to use that, at great enough distances (usually referred to as "at infinity"), the various fields and hence the various potentials are zero; this is an assumption that we have not made explicit until now.

From our earlier work on the existence of potential functions, we know that there is a scalar potential function ϕ such that

$$E = -\nabla \cdot \phi - \frac{\partial A}{\partial t}.$$

But for our E, there are no charges. Thus

$$0 = \nabla \cdot E$$
$$= -\nabla \cdot \left(\nabla \cdot \phi + \frac{\partial A}{\partial t} \right)$$
$$= -\nabla^2 \cdot \phi - \frac{\partial \nabla \cdot A}{\partial t}$$
$$= -\nabla^2 \cdot \phi.$$

Thus we must have that everywhere in space

$$\nabla^2 \cdot \phi = \frac{\partial^2 \phi}{\partial x^2} + \frac{\partial^2 \phi}{\partial y^2} + \frac{\partial^2 \phi}{\partial z^2} = 0.$$

Such functions ϕ are called *harmonic*. Since at the boundary (at "infinity") we are assuming that $\phi = 0$, we get $\phi = 0$ everywhere, as desired. (Again, this results from knowing how to find harmonic functions; for background see chapter 2, section A in Folland's *Introduction to Partial Differential Equations* [22].)

☐

In the preceding proof that $\nabla \cdot A = 0$, we do not use the assumption that E and B satisfy Maxwell's equations when there are no charges or current, just that they satisfy Maxwell's equations in general. Hence we can always choose the vector potential A to be Coulomb. To get a zero function scalar potential, though, we needed to use that there is no charge or current.

If the vector potential A is Coulomb, then it will satisfy a wave equation:

Theorem 15.2.2. *Let A be a Coulomb vector potential for electric and magnetic fields satisfying Maxwell's equations in a free field. Then A satisfies the wave equation*

$$\nabla^2 A - \frac{\partial^2 A}{\partial t^2} = 0.$$

Proof. For ease of notation, we now write $A = (A^1, A^2, A^3)$, instead of our earlier (A_1, A_2, A_3)

Recall that $\nabla^2 A$ denotes

$$\left(\frac{\partial^2 A^1}{\partial x^2} + \frac{\partial^2 A^1}{\partial y^2} + \frac{\partial^2 A^1}{\partial z^2}, \frac{\partial^2 A^2}{\partial x^2} + \frac{\partial^2 A^2}{\partial y^2} + \frac{\partial^2 A^2}{\partial z^2}, \frac{\partial^2 A^3}{\partial x^2} + \frac{\partial^2 A^3}{\partial y^2} + \frac{\partial^2 A^3}{\partial z^2} \right)$$

for the vector field $A = (A^1, A^2, A^3)$. Thus we want to show for each i that

$$\frac{\partial^2 A^i}{\partial x^2} + \frac{\partial^2 A^i}{\partial y^2} + \frac{\partial^2 A^i}{\partial z^2} = -\frac{\partial^2 A^i}{\partial t^2}.$$

Since we are assuming that the scalar potential ϕ is identically zero, we have

$$E = -\frac{\partial A}{\partial t}.$$

Then we have

$$\frac{\partial^2 A}{\partial t^2} = -\partial E/\partial t$$

$$= -\nabla \times B, \text{ by Maxwell}$$

$$= -\nabla \times (\nabla \times A), \text{ since } B = \nabla \times A.$$

Now to calculate $\nabla \times (\nabla \times A)$. To ease notation, we will write

$$\frac{\partial A^i}{\partial x} = A^i_x, \frac{\partial A^i}{\partial y} = A^i_y, \frac{\partial A^i}{\partial z} = A^i_z, \frac{\partial^2 A^i}{\partial x \partial y} = A^i_{xy}, \text{etc.}$$

We have

$$\nabla \times (\nabla \times A) = \nabla \times \det \begin{pmatrix} i & j & k \\ \dfrac{\partial}{\partial x} & \dfrac{\partial}{\partial y} & \dfrac{\partial}{\partial z} \\ A^1 & A^2 & A^3 \end{pmatrix}$$

$$= \nabla \times (A^3_y - A^2_z, A^1_z - A^3_x, A^2_x - A^1_y),$$

which is

$$(A^2_{xy} - A^1_{yy} - A^1_{zz} + A^3_{xz}, A^3_{zy} - A^2_{zz} - A^2_{xx} + A^1_{xy}, A^1_{xz} - A^3_{xx} - A^3_{yy} + A^2_{yz}).$$

Here we will finally use that A is a Coulomb gauge:

$$\nabla \cdot A = A^1_x + A^2_y + A^3_z = 0.$$

Thus

$$A^2_{xy} + A^3_{xz} = \frac{\partial}{\partial x}(A^2_y + A^3_z) = -A^1_{xx}$$

$$A^1_{xy} + A^3_{yz} = \frac{\partial}{\partial y}(A^1_x + A^3_z) = -A^2_{yy}$$

$$A^1_{xz} + A^2_{yz} = \frac{\partial}{\partial z}(A^1_x + A^2_y) = -A^3_{zz},$$

giving us our desired $\nabla^2 A - \frac{\partial^2 A}{\partial t^2} = 0$.

$\square$

The wave equation has been extensively studied, and its solutions are well known. The standard approach is via separation of variables. Almost all books on differential equations will discuss separation of variables, ranging from beginning texts, such as the one by Boyce and DiPrima [6], to graduate level texts, such as the one by Evans [20] and all the way to research monographs, such as Cain and Mayer's *Separation of Variables for Partial Differential Equations* [9]. This allows us just to write down the solutions.

The method of separation of variables starts with the assumption that all solutions to $\nabla^2 A - \frac{\partial^2 A}{\partial t^2} = 0$ are (possibly infinite) linear combinations of solutions of the form

$$f(t)G(x, y, z),$$

and hence a combination of a product of a real-valued function $f(t)$ with a vector-valued function $G(x,y,z)$. Let us assume for a moment that our solution to the wave equation is $A(x,y,z,t) = f(t)G(x,y,z)$. Then we have

$$0 = \nabla^2 A - \frac{\partial^2 A}{\partial t^2}$$

$$= \nabla^2(f(t)G(x,y,z)) - \frac{\partial^2}{\partial t^2}(f(t)G(x,y,z))$$

$$= f(t)\nabla^2 G(x,y,z) - \left(\frac{\partial^2 f}{\partial t^2}\right)G(x,y,z).$$

Writing

$$G(x,y,z) = (G_1(x,y,z), G_2(x,y,z), G_3(x,y,z)),$$

we have

$$\frac{\nabla^2 G_i(x,y,z)}{G_i(x,y,z)} = \frac{\left(\frac{\partial^2 f}{\partial t^2}\right)}{f(t)}$$

for each i. Here is why this technique is called "separation of variables." The left-hand side of the preceding equation is a function of x,y,z alone, while the right-hand side is a function of just the variable t. But this means that each side must be equal to some constant k. Thus we must have G and f satisfying

$$\nabla^2 G(x,y,z) = kG(x,y,z)$$

$$\frac{\partial^2 f}{\partial t^2} = kf(t).$$

To solve only the wave equation, any constant k will work. As seen in the preceding references, though, for the solutions to be periodic, and thus to be what we would want to consider as waves, we need k to be negative; that is why we can set

$$k = -\omega^2,$$

for a constant ω. Thus we want to find solutions to

$$\nabla^2 G(x,y,z) = -\omega^2 G(x,y,z)$$

$$\frac{\partial^2 f}{\partial t^2} = -\omega^2 f(t).$$

The solution $G(x,y,z)$ is a vector field with x-coordinate

$$a\sin(k_1\omega x + k_2\omega y + k_3\omega z + \alpha),$$

y-coordinate

$$b\sin(k_1\omega x + k_2\omega y + k_3\omega z + \alpha),$$

and z-coordinate

$$c \sin(k_1 \omega x + k_2 \omega y + k_3 \omega z + \alpha),$$

where (k_1, k_2, k_3) is a vector of length one, and the phase α and the coefficients $a, b,$ and c are real constants. The solution $f(t)$ is of the form

$$f(t) = \sin(\omega t + \beta),$$

where the phase β is a real constant. To ease notation, we will assume that the phases α and β are both zero. Also to ease notation, we will reorient our x, y, z coordinate system so that $(k_1, k_2, k_3) = (0, 0, 1)$. None of these assumptions fundamentally alters the underlying mathematics. We then have

$$G(x, y, z) = (a \sin(\omega z), b \sin(\omega z), c \sin(\omega z))$$

$$f(t) = \sin(\omega t).$$

We know that

$$\nabla \cdot A = 0.$$

Then

$$0 = \frac{\partial(\sin(\omega t)a \sin(\omega z))}{\partial x} + \frac{\partial(\sin(\omega t)b \sin(\omega z))}{\partial y} + \frac{\partial(\sin(\omega t)c \sin(\omega z))}{\partial z}$$

$$= c \sin(\omega t)\cos(\omega z),$$

forcing $c = 0$, meaning

$$A(x, y, z, t) = (a \sin(\omega t)\sin(\omega z), b \sin(\omega t)\sin(\omega z), 0).$$

Then the electric and magnetic fields are

$$E(z, t) = -\frac{\partial A}{\partial t}$$

$$= -\frac{\partial f(t)}{\partial t} G(x, y, z)$$

$$= -\omega \cos(\omega t)G(x, y, z)$$

$$= (-a\omega \cos(\omega t)\sin(\omega z), -b\omega \cos(\omega t)\sin(\omega z), 0)$$

$$B(z, t) = \nabla \times A$$

$$= f(t)\nabla \times G(x, y, z)$$

$$= f(t)\nabla \times (a \sin(\omega z), b \sin(\omega z), 0)$$

$$= (-b\omega \sin(\omega t)\cos(\omega z), a\omega \sin(\omega t)\cos(\omega z), 0).$$

15.3. The "Hidden" Harmonic Oscillator

We are close to our goal of showing that the Hamiltonian for a monochromatic light wave has the same form as a harmonic oscillator's Hamiltonian.

For a physical wave, we know that the Hamiltonian (the energy) is proportional to the square of the wave's amplitude. We want to make the claim that the energy density for an electromagnetic wave is

$$E \cdot E + B \cdot B,$$

in analogy to the square of the amplitude of a traditional wave.

Here is the intuitive justification. (Of course, this formula for the energy of an electromagnetic wave can be checked experimentally.) An electromagnetic wave is made up of two separate waves, namely, the electric wave and the magnetic wave. The energy density of the electric wave will be the square $E \cdot E$ of its amplitude E, and the energy density of the magnetic wave will be the square $B \cdot B$ of its amplitude B. The total energy should simply be the sum of these two energies, giving us our desired energy density of $E \cdot E + B \cdot B$.

To find the Hamiltonian, we need to integrate this energy density over an appropriate volume V. Using the notation for A, E, and B from the previous section, we have the constant ω. We choose for our volume a box with volume V whose sides are parallel to the x-y plane, x-z plane, and y-z plane, with side lengths of $\sqrt{\omega/\pi}$ in the directions of the x- and y-axes and $2\pi/\omega$ in the direction of the z-axis. The choice of $2\pi/\omega$ for the z-axis is natural, reflecting how we set-up the electric and magnetic fields in the previous section. The choice of $\sqrt{\omega/\pi}$ for the x- and y-axes is a bit more arbitrary but will make some of the integrals that follow turn out cleanly.

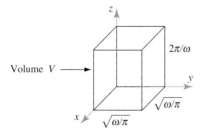

Figure 15.1

We will use that

$$\int_0^{\frac{2\pi}{\omega}} \sin^2(\omega z) \, dz = \int_0^{\frac{2\pi}{\omega}} \cos^2(\omega z) \, dz = \frac{\pi}{\omega}.$$

Then the Hamiltonian over the volume V will be

$$H(t) = \int_V (E \cdot E + B \cdot B)\, dx\, dy\, dz$$

$$= \int_V a^2 \omega^2 \cos^2(\omega t) \sin^2(\omega z)\, dx\, dy\, dz$$

$$+ \int_V b^2 \omega^2 \cos^2(\omega t) \sin^2(\omega z)\, dx\, dy\, dz$$

$$+ \int_V b^2 \omega^2 \sin^2(\omega t) \cos^2(\omega z)\, dx\, dy\, dz$$

$$+ \int_V a^2 \omega^2 \sin^2(\omega t) \cos^2(\omega z)\, dx\, dy\, dz$$

$$= a^2 \omega^2 \cos^2(\omega t) + b^2 \omega^2 \cos^2(\omega t)$$

$$+ b^2 \omega^2 \sin^2(\omega t) + a^2 \omega^2 \sin^2(\omega t).$$

Of course, we could rewrite this as simply $\omega^2(a^2 + b^2) = \omega^2$, since $a^2 + b^2$ was assumed to be 1, which reflects that the Hamiltonian for a monochromatic wave is a constant. We will resist this temptation, as our goal is not the cleanest formula for the classical electromagnetic Hamiltonian, but instead a formulation that looks like the Hamiltonian for a harmonic oscillator, allowing us to quantize.

Still following the outline in [41], we set

$$\alpha(t) = e^{-i\omega t},$$

giving us that

$$\sin(\omega t) = \frac{i}{2}(\alpha(t) - \bar{\alpha}(t))$$

$$\cos(\omega t) = \frac{1}{2}(\alpha(t) + \bar{\alpha}(t)).$$

Then set

$$q(t) = \frac{i}{2}(\alpha(t) - \bar{\alpha}(t))$$

$$p(t) = \frac{\omega}{2}(\alpha(t) + \bar{\alpha}(t)).$$

Then the Hamiltonian is

$$H(t) = p^2 + \omega^2 q^2.$$

Here is why we made these seemingly random substitutions. Recall that the Hamiltonian for a harmonic oscillator is

$$H_{\text{harmonic oscillator}} = \frac{1}{2m}p^2 + \frac{m\omega^2}{2}x^2,$$

where now x is the position and p is the momentum. For the electromagnetic Hamiltonian, the function q plays the role of position and the function p the role of the momentum. In order for this analog to hold, note that

$$\frac{dq}{dt} = p, \quad \frac{dp}{dt} = -\omega^2 q,$$

in direct analog to the role that the variables x and p play in the Hamiltonian for the quantum harmonic oscillator.

Thus, mathematically, the Hamiltonian for a monochromatic electromagnetic wave has the same form as the Hamiltonian for a harmonic oscillator. This result will allow us easily to quantize Maxwell's equations in the next section.

15.4. Quantization of Maxwell's Equations

We are finally ready to quantize the electromagnetic fields. We have just seen that the Hamiltonian for the electromagnetic field can be put into the same form as that of the Hamiltonian for a harmonic oscillator. From the last section, the Hamiltonian for a monochromatic electromagnetic wave is

$$H(t) = \frac{1}{2}(p^2 + \omega^2 q^2)$$

where $q(t) = \frac{i}{2}(\alpha(t) - \bar{\alpha}(t))$ and $p(t) = \frac{\omega}{2}(\alpha(t) + \bar{\alpha}(t))$. To quantize, motivated by our work for the harmonic oscillator, we replace the continuous variable $\alpha(t)$ with an appropriate operator.

Let $a(t)$ be an operator with adjoint $a^*(t)$ such that

$$[a, a^*] = 1.$$

Replace the continuous variable $\alpha(t)$ by the operator

$$\sqrt{\frac{\hbar}{\omega}} a(t),$$

and, hence, replace $\bar{\alpha}(t)$ by the operator

$$\sqrt{\frac{\hbar}{\omega}} a^*(t).$$

We will still call $a^*(t)$ and $a(t)$ the creation and annihilation operators.

Then we replace our continuous function $q(t)$ by the operator (still denoted as $q(t)$)

$$q(t) = \frac{i}{2}\sqrt{\frac{\hbar}{\omega}}\left(a(t) - a^*(t)\right)$$

and the continuous function $p(t)$ by the operator (still denoted as $p(t)$)

$$p(t) = \frac{\omega}{2}\sqrt{\frac{\hbar}{\omega}}\left(a(t) + a^*(t)\right).$$

Then, using that $[a, a^*] = 1$ implies $aa^* = a^*a + 1$, the quantized version of the Hamiltonian for the electromagnetic field is

$$H = \hbar\omega\left(a^*a + \frac{1}{2}\right).$$

We know that the eigenvalues of the Hamiltonian are of the form

$$E_n = \left(n + \frac{1}{2}\right)\hbar\omega.$$

We can interpret the integer n as the number of photons of frequency ω. Note that, even when $n = 0$, there is still a non-zero energy. The big news is that our formulation has led to the prediction of photons, allowing us at last to start explaining the photoelectric effect.

We now want to quantize the vector potential $A(x, y, z, t)$, the electric field $E(x, y, z, t)$, and the magnetic field $B(x, y, z, t)$. This procedure is straightforward. We just replace each occurrence of $\alpha(t)$ with the operator $\sqrt{\frac{\hbar}{\omega}}a(t)$ and each $\overline{\alpha}(t)$ with the operator $\sqrt{\frac{\hbar}{\omega}}a^*(t)$. In the classical case we have

$$A(x, y, z, t) = \sin(\omega t)G(x, y, z) = \frac{i}{2}\left(\alpha(t) - \overline{\alpha}(t)\right)G(x, y, z),$$

where G is a vector field. Then the quantized version is

$$A = \frac{i}{2}\left(\sqrt{\frac{\hbar}{\omega}}a(t) - \sqrt{\frac{\hbar}{\omega}}a^*(t)\right)G(x, y, z),$$

with G still a vector field.

Similarly, as we saw in Section 15.2, the classical E and B are

$$E(x, y, z, t) = -\omega\cos(\omega t)G(x, y, z) = \frac{-\omega}{2}\left(\alpha(t) + \overline{\alpha}(t)\right)G(x, y, z)$$

and

$$B(x, y, z, t) = \sin(\omega t)(\nabla \times G(x, y, z)) = \frac{i}{2}\left(\alpha(t) - \overline{\alpha}(t)\right)(\nabla \times G(x, y, z)).$$

Then the quantized versions must be

$$E = \frac{-\omega}{2} \left(\sqrt{\frac{\hbar}{\omega}} a(t) + \sqrt{\frac{\hbar}{\omega}} a^*(t) \right) G(x,y,z)$$

and

$$B = \frac{i}{2} \left(\sqrt{\frac{\hbar}{\omega}} a(t) - \sqrt{\frac{\hbar}{\omega}} a^*(t) \right) (\nabla \times G(x,y,z)).$$

15.5. Exercises

Exercise 15.5.1. *Let* $F(x)$ *be a function of* x *and* $G(t)$ *be a function of a different variable* t. *Suppose for all* x *and* t *that*

$$F(x) = G(t).$$

Show that there is a constant α *such that for all* x

$$F(x) = \alpha$$

and for all t

$$G(t) = \alpha.$$

Exercise 15.5.2. *Suppose that a solution* $f(x,t)$ *to the partial differential equation*

$$\frac{\partial^2 f}{\partial x^2} = \frac{\partial^2 f}{\partial t^2}$$

can be written as the product of two one-variable functions $X(x)$ *and* $T(t)$:

$$f(x,t) = X(x)T(t).$$

Using the first problem, show that there is a constant α *such that* $X(x)$ *satisfies the ordinary differential equation*

$$\frac{d^2 X(x)}{dx^2} = \alpha X(x)$$

and $T(t)$ *satisfies the ordinary differential equation*

$$\frac{d^2 T(t)}{dt^2} = \alpha T(t).$$

Exercise 15.5.3. *Show that the product*

$$A(r,t) = \alpha(t)A_0(r)$$

satisfies the wave equation

$$\nabla^2(A(r,t)) - \frac{\partial^2(A(r,t))}{\partial t^2} = 0,$$

if $A_0(r)$ is a vector-valued function, where r denotes the position variables (x,y,z), satisfying

$$\nabla^2 A_0(r) + k^2 A_0(r) = 0,$$

and $\alpha(t)$ is a real-valued function satisfying

$$\frac{d^2\alpha(t)}{dt^2} = -k^2\alpha(t),$$

and k is any constant.

Exercise 15.5.4. *Show that the vector field*

$$E(x,y,x,t) = (E_1(x,y,z,t), E_2(x,y,z,t), E_3(x,y,z,t))$$

where

$$E_i(x,y,z,t) = E_i \sin(k_1\omega x + k_2\omega y + k_3\omega z - \omega t),$$

with each E_i a constant, the vector (k_1,k_2,k_3) having length one and ω being a constant, satisfies the wave equation

$$\nabla^2 E - \frac{\partial^2 E}{\partial t^2} = 0.$$

Exercise 15.5.5. *Let*

$$E = (E_1(\sin(\omega z - \omega t), E_2(\sin(\omega z - \omega t), 0)$$

and

$$B = (E_2\cos(\omega z - \omega t), -E_1\cos(\omega z - \omega t), 0).$$

Show that

$$E \cdot E + B \cdot B = E_1^2 + E_2^2$$

and

$$E \cdot B = 0.$$

Exercise 15.5.6. *Let*

$$\alpha(t) = \alpha(0)e^{-i\omega t}, \ \overline{\alpha}(t) = \overline{\alpha}(0)e^{i\omega t}$$

be solutions to $\frac{d^2\alpha(t)}{dt^2} = -\omega^2\alpha(t)$. Letting $q(t) = \frac{i}{2}(\alpha(t) - \overline{\alpha}(t))$ and $p(t) = \frac{\omega}{2}(\alpha(t) + \overline{\alpha}(t))$, show that

$$\frac{dq}{dt} = p, \frac{dp}{dt} = -\omega^2 q.$$

Exercise 15.5.7. *Show*

$$\int_0^{\frac{2\pi}{\omega}} \sin^2(\omega z)\,dz = \int_0^{\frac{2\pi}{\omega}} \cos^2(\omega z)\,dz = \frac{\pi}{\omega}.$$

Exercise 15.5.8. *Using the notation from Section 15.3, show that*

$$\sin(\omega t) = \frac{i}{2}(\alpha(t) - \overline{\alpha}(t))$$

and

$$\cos(\omega t) = \frac{1}{2}(\alpha(t) + \overline{\alpha}(t)).$$

Exercise 15.5.9. *Using the notation of Section 15.3, show that*

$$H(t) = p^2 + \omega^2 q^2.$$

Exercise 15.5.10. *Using the notation from Section 15.4, show that*

$$[a(t), a^*(t)] = 1$$

implies

$$[q(t), p(t)] = i\hbar.$$

Exercise 15.5.11. *Let $A : V \to V$ be any linear operator from a Hilbert space V to itself. Show that*

$$A + A^*$$

must be Hermitian. (You may assume that $A^{**} = A$, as is indeed always the case.)

Exercise 15.5.12. *Let $A : V \to V$ be any linear operator from a Hilbert space V to itself. Show that*

$$i(A - A^*)$$

must be Hermitian.

Exercise 15.5.13. *Using the notation from Section 15.4, show that $q(t)$, $p(t)$, and $H(t)$ are all Hermitian.*

Exercise 15.5.14. *Using the notation from 15.4, show that the operator corresponding to the classical electric field,*

$$E = \frac{-\omega}{2}\left(\sqrt{\frac{\hbar}{\omega}}a(t) + \sqrt{\frac{\hbar}{\omega}}a^*(t)\right)G(x,y,z),$$

is Hermitian.

Exercise 15.5.15. *Using the notation from 15.4, show that the operator corresponding to the classical magnetic field*

$$B = \frac{i}{2} \left(\sqrt{\frac{\hbar}{\omega}} a(t) - \sqrt{\frac{\hbar}{\omega}} a^*(t) \right) \nabla \times G(x,y,z)$$

is Hermitian.

16

Manifolds

Summary: The goal of this chapter is to introduce manifolds, which are key objects for geometry. We will give three different ways for defining manifolds: parametrically, implicitly, and abstractly.

16.1. Introduction to Manifolds

16.1.1. Force = Curvature

The goal for the rest of this book is to understand the idea that

$$\text{Force} = \text{Curvature}.$$

This idea is one of the great insights in recent science. In this chapter we introduce manifolds, which are basically the way to think about decent geometric objects. In the next chapter we will introduce vector bundles on manifolds. In the following chapter, we will introduce connections, which are fancy ways of taking derivatives. In Chapter 19, we will define different notions for curvature. Finally, in Chapter 20, in the context of Maxwell's equations, we will see how "Force = Curvature."

16.1.2. Intuitions behind Manifolds

(This section is really the same as the first few paragraphs of section 6.4 in [26].)

While manifolds are, to some extent, some of the most naturally occurring geometric objects, it takes work and care to create correct definitions. In essence, a k-dimensional manifold is any space that, in a neighborhood of any point, looks like a ball in $\mathbb{R}^k$. We will be concerned at first with manifolds that live in some ambient $\mathbb{R}^n$. For this type of manifold, we give two equivalent definitions: the parametric version and the implicit version. For each of these versions, we will carefully show that the unit circle S^1

Figure 16.1

in $\mathbb{R}^2$ is a one-dimensional manifold. (Of course, if we were just interested in circles we would not need all of these definitions; we are just using the circle to get a feel for the correctness of the definitions.) Then we will define an abstract manifold, a type of geometric object that need not be defined in terms of some ambient $\mathbb{R}^n$.

Consider the circle S^1. Near any point $p \in S^1$ the circle looks like an interval (admittedly a bent interval). In a similar fashion, we want our definitions to yield that the unit sphere S^2 in $\mathbb{R}^3$ is a two-dimensional manifold, since near any point $p \in S^2$,

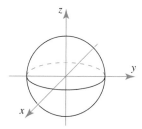

Figure 16.2

the sphere looks like a disc (though, again, more like a bent disc). We want to exclude from our definition of manifold objects that contain points for which there is no well-defined notion of a tangent space, such as the cone:

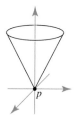

Figure 16.3

which has tangent difficulties at the vertex p.

16.2. Manifolds Embedded in $\mathbb{R}^n$

16.2.1. Parametric Manifolds

(This is an expansion of Section 9.2.)

Let M be a set of points in $\mathbb{R}^n$. Intuitively we want to say that M is a k-dimensional manifold, if, for any point p on M, M looks, up close and near p, like $\mathbb{R}^k$.

Before giving the definition, we need some notation. Let $x = (x_1, x_2, \ldots, x_n)$ be the coordinates for $\mathbb{R}^n$ and let $u = (u_1, \ldots, u_k)$ be coordinates in $\mathbb{R}^k$. Let $k \leq n$ and let U be an open ball in $\mathbb{R}^k$ centered at the origin.

A map

$$\gamma : U \to \mathbb{R}^n$$

is given by specifying n functions $x_1(u), \ldots, x_n(u)$ such that

$$\begin{aligned}
\gamma(u) &= \gamma(u_1, u_2, \ldots, u_k) \\
&= (x_1(u_1, \ldots, u_k), \ldots, x_n(u_1, \ldots, u_k)) \\
&= (x_1(u), \ldots, x_n(u)).
\end{aligned}$$

The map γ is said to be *continuous* if the functions $x_1(u), \ldots, x_n(u)$ are continuous, *differentiable* if the functions are differentiable, and so on. We will assume that our map γ is differentiable. The Jacobian of the map γ is the $n \times k$ matrix

$$J(\gamma) = \begin{pmatrix} \frac{\partial x_1}{\partial u_1} & \cdots & \frac{\partial x_1}{\partial u_k} \\ & \vdots & \\ \frac{\partial x_n}{\partial u_1} & \cdots & \frac{\partial x_n}{\partial u_k} \end{pmatrix}.$$

We say that the rank of the Jacobian is k if the preceding matrix has a rank $k \times k$ submatrix or, in other words, if one of the $k \times k$ minors is an invertible matrix. We can now define (again) a k-dimensional manifold in $\mathbb{R}^n$.

Definition 16.2.1. *A set of points M in $\mathbb{R}^n$ is a k-dimensional manifold if for every point $p \in M$, there exist a small open ball V in $\mathbb{R}^n$, centered at p, and a differentiable map*

$$\gamma : U \to \mathbb{R}^n,$$

with U an open ball in $\mathbb{R}^k$, centered at the origin, such that

1. *$\gamma(0, \ldots, 0) = p$,*
2. *γ is a one-to-one onto map from U to $M \cap V$,*
3. *The rank of the Jacobian for γ is k at every point.*

Let us look at an example: We show that the circle S^1 is indeed a one-dimensional manifold via the preceding definition, for the point $p = (1,0)$.

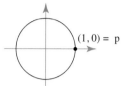

Figure 16.4

Our map will be

$$\gamma(t) = (\sqrt{1-t^2}, t).$$

Note that

$$\gamma(0) = (1,0).$$

Let our open "ball" U be the open interval $\{t : -1/2 < t < 1/2\}$. To find the corresponding open ball V in $\mathbb{R}^2$, let the radius of V be

$$r = \sqrt{\left(\frac{\sqrt{3}}{2} - 1\right)^2 + \frac{1}{4}}.$$

(This r is just the distance from the point $\gamma(1/2) = (\sqrt{3}/2, 1/2)$ to $\gamma(0) = (1,0)$.)

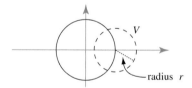

Figure 16.5

It is not hard to show that γ is indeed a one-to-one onto map from U to $S^1 \cap V$. We have the Jacobian being

$$J(\gamma) = \left(\frac{\partial \sqrt{1-t^2}}{\partial t}, \frac{\partial t}{\partial t}\right) = \left(\frac{-t}{\sqrt{1-t^2}}, 1\right).$$

Though the function $-t/\sqrt{1-t^2}$ is not defined at $t = \pm 1$, this is no problem, as we chose our open interval U to be strictly smaller than $[-1,1]$. Since the second coordinate of the Jacobian is the constant function 1, we know that the Jacobian has rank one.

To show that the circle is a one-dimensional manifold at the point $(0, 1)$, we would need to use a different parameterizing map, such as

$$\gamma(t) = (t, \sqrt{1 - t^2}).$$

16.2.2. Implicitly Defined Manifolds

(This is repeating much of section 6.4 in [26].)

We could have also described the circle S^1 as

$$S^1 = \{(x, y) : x^2 + y^2 - 1 = 0\}.$$

If we set

$$\rho(x, y) = x^2 + y^2 - 1,$$

then

$$S^1 = \{(x, y) : \rho(x, y) = 0\}.$$

We would then say that the circle S^1 is the zero locus of the function $\rho(x, y)$.

This suggests a totally different way for defining manifolds, namely, as zero loci of a set of functions on $\mathbb{R}^n$.

Definition 16.2.2 (Implicit Manifolds). *A set M in $\mathbb{R}^n$ is a k-dimensional manifold if for any point $p \in M$ there is an open set U containing p and $(n - k)$ differentiable functions $\rho_1, \ldots, \rho_{n-k}$ such that*

1. $M \cap U = (\rho_1 = 0) \cap \cdots \cap (\rho_{n-k} = 0)$,
2. At all points in $M \cap U$, the gradient vectors

$$\nabla \rho_1, \ldots, \nabla \rho_{n-k}$$

are linearly independent.

(From multivariable calculus, the vectors $\nabla \rho_1, \ldots, \nabla \rho_{n-k}$ are all normal vectors.) Intuitively, each function ρ_i cuts down the degree of freedom by one. If we have $n - k$ functions on $\mathbb{R}^n$, then we are left with k degrees of freedom, giving some sense as to why this should also be a definition for k-dimensional manifolds.

Returning to the circle, for $\rho = x^2 + y^2 - 1$ we have

$$\nabla(x^2 + y^2 - 1) = (2x, 2y),$$

which is never the zero vector at any point on the circle.

16.3. Abstract Manifolds

16.3.1. Definition

It took mathematicians many years to develop the definition of, or even to realize the need for, abstract manifolds. Here we want to have a geometric object that is independent of being defined in an ambient $\mathbb{R}^n$.

We will be following closely the definition of a manifold given in section 1.1 of [49].

Definition 16.3.1. *A set of points M is a smooth n-dimensional manifold if there are a countable collection $\mathcal{U} = \{U_\alpha\}$ of subsets of M, called the coordinate charts; a countable collection $\mathcal{V} = \{V_\alpha\}$ of connected open subsets of $\mathbb{R}^n$; and one-to-one onto maps $\chi_\alpha : V_\alpha \to U_\alpha$, such that*

1. The coordinate charts U_α cover M, meaning that

$$\bigcup U_\alpha = M.$$

2. If for two coordinate charts U_α and U_β we have non-empty intersection

$$U_\alpha \cap U_\beta \neq \emptyset,$$

   then the map

$$\chi_\beta^{-1} \circ \chi_\alpha : \chi_\alpha^{-1}(U_\alpha \cap U_\beta) \to \chi_\beta^{-1}(U_\alpha \cap U_\beta)$$

   is differentiable.

3. Let $x \in U_\alpha$ and $y \in U_\beta$ be two distinct points on M. Then there exist open subsets $W \subset V_\alpha$ and $W' \subset V_\beta$ with

$$\chi_\alpha^{-1}(x) \in W \quad and \quad \chi_\beta^{-1}(y) \in W'$$

   and

$$\chi_\alpha(W) \cap \chi_\beta(W') = \emptyset.$$

We say that $(\mathcal{U}, \mathcal{V}, \{\chi_\alpha\})$ defines a manifold structure on M.

By only requiring that the various maps $\chi_\beta^{-1} \circ \chi_\alpha$ be continuous, we have the definition for a topological manifold. Similarly, if the maps $\chi_\beta^{-1} \circ \chi_\alpha$ are real-analytic, we have real-analytic manifolds, and so on.

We now have to unravel the meaning of this definition. Underlying this definition is that we know all about maps $F : \mathbb{R}^n \to \mathbb{R}^n$. Such F are defined by setting

$$F(x_1, \ldots, x_n) = (f_1(x_1, \ldots, x_n), \ldots, f_n(x_1, \ldots, x_n)),$$

where each $f_i : \mathbb{R}^n \to \mathbb{R}$. The map F is differentiable if all first-order derivatives for each f_i exist.

We want to think of each coordinate chart U_α as "carrying" part of $\mathbb{R}^n$.

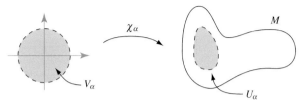

Figure 16.6

The condition $\bigcup U_\alpha = M$ is simply stating that each point of M is in at least one of the coordinate charts.

Condition 2 is key. We have

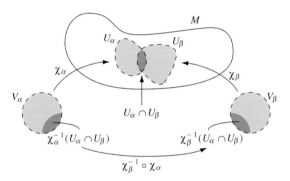

Figure 16.7

Then each $\chi_\beta^{-1} \circ \chi_\alpha$ is a map from $\mathbb{R}^n$ to $\mathbb{R}^n$. We can describe these maps' differentiability properties.

The third condition states that we can separate points in M. For those who know topology, this condition is simply giving us that the set M is Hausdorff.

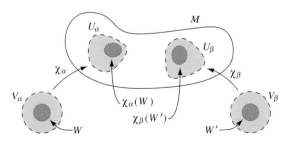

Figure 16.8

Let us look at a simple example, the unit circle $S^1 = \{(x,y) : x^2 + y^2 = 1\}$. We set

$$U_1 = \{(x,y) \in S^1 : x > 0\}$$

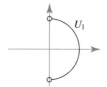

Figure 16.9

$$U_2 = \{(x,y) \in S^1 : y > 0\}$$

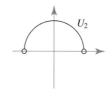

Figure 16.10

$$U_3 = \{(x,y) \in S^1 : x < 0\}$$

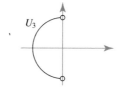

Figure 16.11

$$U_4 = \{(x,y) \in S^1 : y < 0\}$$

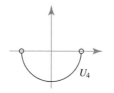

Figure 16.12

For all four charts, we have

$$V_1 = V_2 = V_3 = V_4 = \{t \in \mathbb{R} : -1 < t < 1\}.$$

We set

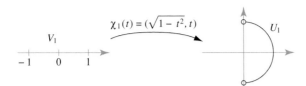

Figure 16.13

Then

$$\chi_1^{-1}(x,y) = y.$$

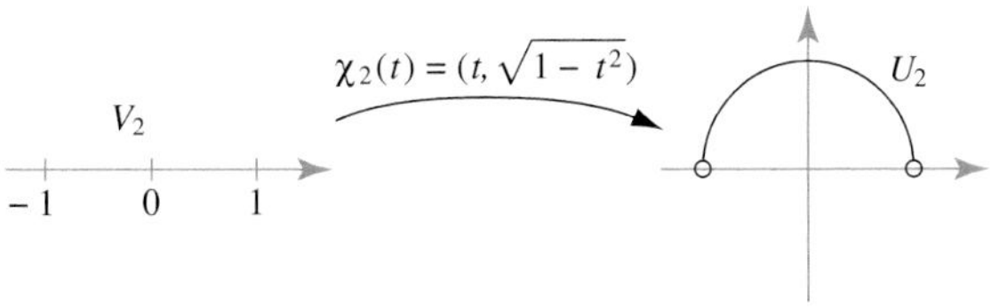

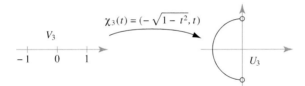

Figure 16.14

$$\chi_2^{-1}(x,y) = x.$$

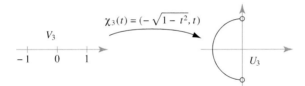

Figure 16.15

$$\chi_3^{-1}(x,y) = y.$$

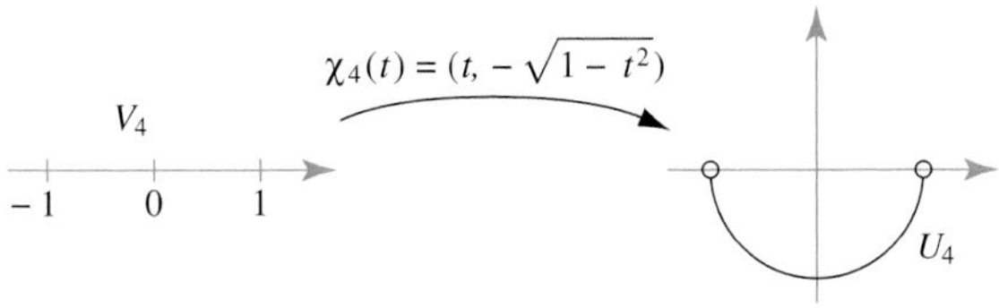

Figure 16.16

$$\chi_4^{-1}(x,y) = x.$$

Let us look at the overlaps. We have that

$$U_1 \cap U_2 = \{(x,y) \in S^1 : x > 0, y > 0\}$$

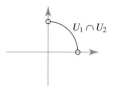

Figure 16.17

with

$$\chi_1^{-1}(U_1 \cap U_2) = \{t \in \mathbb{R} : 0 < t < 1\} = \chi_2^{-1}(U_1 \cap U_2).$$

Then

$$\chi_2^{-1} \circ \chi_1(t) = \chi_2^{-1}(\sqrt{1-t^2}, t) = \sqrt{1-t^2}.$$

Similarly, we get

$$\chi_3^{-1} \circ \chi_2(t) = \chi_3^{-1}(t, \sqrt{1-t^2}) = \sqrt{1-t^2}$$

$$\chi_4^{-1} \circ \chi_3(t) = \chi_4^{-1}(-\sqrt{1-t^2}, t) = -\sqrt{1-t^2}$$

$$\chi_1^{-1} \circ \chi_4(t) = \chi_4^{-1}(t, -\sqrt{1-t^2}) = -\sqrt{1-t^2}.$$

All of these maps are infinitely differentiable real-valued maps.

Now for a technicality. The proceeding depends on choosing coordinate charts. For example, we could have started with our circle S^1 and chosen coordinate charts

$$U_1' = \left\{ (x, y) \in S^2 : -\frac{\sqrt{3}}{2} < x < \frac{\sqrt{3}}{2}, y > 0 \right\}$$

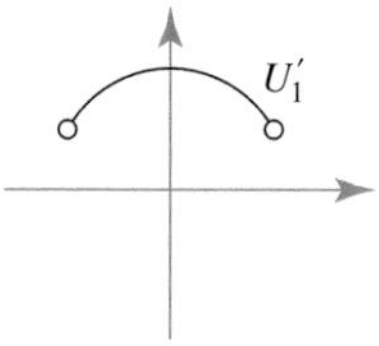

Figure 16.18

$$U_2' = \left\{ (x, y) \in S^2 : -\frac{\sqrt{3}}{2} < y < \frac{\sqrt{3}}{2}, x < 0 \right\}$$

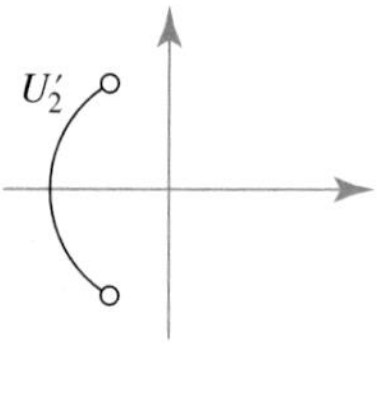

Figure 16.19

$$U_3' = \left\{ (x,y) \in S^2 : -\frac{\sqrt{3}}{2} < x < \frac{\sqrt{3}}{2}, y < 0 \right\}$$

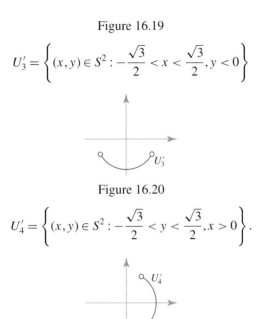

Figure 16.20

$$U_4' = \left\{ (x,y) \in S^2 : -\frac{\sqrt{3}}{2} < y < \frac{\sqrt{3}}{2}, x > 0 \right\}.$$

Figure 16.21

Surely we want to say that this is the same manifold. We need to be able somehow to compare different coordinate charts for the same set of points M. Let $\mathcal{U} = \{U_\alpha\}$ be a collection of coordinate charts on M, with $\mathcal{V} = \{V_\alpha\}$ the corresponding connected open sets in $\mathbb{R}^n$ and $\chi_\alpha : V_\alpha \to U_\alpha$ the maps that make M into a manifold. Now consider another collection of coordinate charts, denoted by $\mathcal{U}' = \{U_\mu'\}$, with corresponding connected open sets $\mathcal{V} = \{V_\mu'\}$ in $\mathbb{R}^n$ and maps $\chi_\mu' : U_\mu' \to U_\mu'$. We want to know when $(\mathcal{U}, \mathcal{V}, \{\chi_\alpha\})$ and $(\mathcal{U}', \mathcal{V}', \{\chi_\mu'\})$ define the "same" manifold.

Definition 16.3.2. *We say that $(\mathcal{U}, \mathcal{V}, \{\chi_\alpha\})$ and $(\mathcal{U}', \mathcal{V}', \{\chi_\mu'\})$ are compatible if the maps*

$$\chi_\mu'^{-1} \circ \chi_\alpha : \chi_\alpha^{-1}(U_\alpha \cap U_\mu') \to \chi_\mu'^{-1}(U_\alpha \cap U_\mu')$$

are differentiable, in which case we say they define the same manifold structure on M.

Finally, this definition for a manifold allows us to have a natural notion of open sets.

Definition 16.3.3. *A subset U of a manifold M is open if for all α the sets*

$$\chi_\alpha(U \cap U_\alpha)$$

are open sets of $\mathbb{R}^n$.

16.3.2. Functions on a Manifold

We can now talk about what it means for a function to be differentiable on a manifold. Again, we will reduce the definition to a statement about the differentiability of a function from $\mathbb{R}^n$ to $\mathbb{R}$.

Definition 16.3.4. *A real-valued function f on a manifold M is differentiable if, for an open cover (U_α) and maps ϕ_α : Open ball in $\mathbb{R}^n \to U_\alpha$, the composition functions*

$$f \circ \phi_\alpha : \text{Open ball in } \mathbb{R}^n \to \mathbb{R}$$

are differentiable, for all α.

16.4. Exercises

Exercise 16.4.1. *In $\mathbb{R}^3$, the unit sphere S^2 is the set of points distance one from the origin and thus the points $(x, y, z) \in \mathbb{R}^3$ such that*

$$x^2 + y^2 + z^2 = 1.$$

Using the definition of implicitly defined manifold, show that S^2 is a manifold.

Exercise 16.4.2. *Using the definition for parametrically defined manifolds, show that S^2 is a manifold.*

Exercise 16.4.3. *Using the definition for abstractly defined manifolds, show that S^2 is a manifold.*

Exercise 16.4.4. *Show that*

$$\{(x, y, z) \in \mathbb{R}^3 : x^2 + y^2 + z^2 = 1, x + y + z = 0\}$$

is an implicit manifold of dimension one. What type of geometric object is it?

Exercise 16.4.5. *Show that $\{(x, y, z) \in \mathbb{R}^3 : x^2 + y^2 + z^2 = 1, x + y + z = 0\}$ can also be proven to be a parametrically defined dimension one manifold.*

Exercise 16.4.6. *Show that $z^2 = x^2 + y^2$ is not a manifold at the origin.*

Exercise 16.4.7. *Show that in $\mathbb{R}^2$, the zero set*

$$x^2 - y^2 = 0$$

is not a manifold.

Exercise 16.4.8. *Let* M *be an abstract manifold with covering* $\{U_\alpha\}$*. Let* U *and* U' *be two open sets on* M*. Prove that* $U \cup U'$ *and* $U \cap U'$ *are also open.*

Exercise 16.4.9. *Show that on the unit circle*

$$S^1 = \{(x,y) \in \mathbb{R}^2 : x^2 + y^2 = 1\},$$

the function

$$f(x,y) = x$$

is differentiable. (Here you need to think in terms of coordinate charts.)

Exercise 16.4.10. *Show that on the unit circle* S^1 *the function*

$$f(x,y) = |x|$$

is continuous but not differentiable at two points.

17

Vector Bundles

Summary: The basics of vector bundles are given in this chapter. As we will see in the next two chapters, vector bundles are needed to understand curvature of manifolds correctly. Further, we will eventually see that vector bundles are needed to generalize Maxwell's equations to other forces.

17.1. Intuitions

Picture a surface in space.

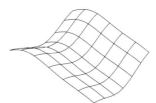

Figure 17.1

In the previous chapter, we saw methods for describing manifolds. But now consider the surface in Figure 17.2.

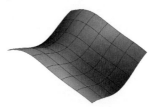

Figure 17.2 (see Plate 1 for color version)

How can we account for the color? One method would be still to describe the surface as a manifold, but add the following type of extra information. At each point of the surface, attach a color to the point. Imagine a color wheel (Figure 17.3).

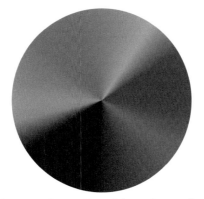

Figure 17.3 (see Plate 2 for color version)

Then corresponding to each point of the surface is a point on the color wheel. We can now describe our colorful surface by points in the product of

$$\text{Colorless surface} \times \text{Color Wheel.}$$

This happens all the time. You have a geometric object that has extra information attached to each point. For example, we can imaging a spinning top moving along a curve in space

Figure 17.4

Here we can attach to each point of the curve the direction about which the top is spinning.

This is the beginnings of the theory of fiber bundles. You start with a manifold and attach to each point extra information. All the possible values for the extra information are called the fiber over the point. For our initial color example, the fiber would be the entire color wheel.

For us, the extra information will be a vector space. We want somehow to attach, at each point of our manifold, a vector space, and to have these attached vector spaces vary continuously as we vary the point on the manifold.

17.2. Technical Definitions

17.2.1. The Vector Space $\mathbb{R}^k$

In this book, the vector spaces making up our vector bundles will all be various real vector spaces $\mathbb{R}^k$. Here we just want to fix some notation. For us, $\mathbb{R}^k$ will be column vectors:

$$\mathbb{R}^k = \left\{ \begin{pmatrix} x_1 \\ \vdots \\ x_k \end{pmatrix} : x_i \in \mathbb{R} \right\}.$$

The natural maps from the vector space $\mathbb{R}^k$ to itself are given by matrix multiplications of $k \times k$ matrices times the column vectors of $\mathbb{R}^k$. Thus, given an $n \times n$ matrix

$$g = \begin{pmatrix} g_{11} & \cdots & g_{1k} \\ \vdots & \cdots & \vdots \\ g_{k1} & \cdots & g_{kk} \end{pmatrix},$$

we have the map

$$g : \mathbb{R}^k \to \mathbb{R}^k$$

given by

$$gx = \begin{pmatrix} g_{11} & \cdots & g_{1k} \\ \vdots & \cdots & \vdots \\ g_{k1} & \cdots & g_{kk} \end{pmatrix} \begin{pmatrix} x_1 \\ \vdots \\ x_n \end{pmatrix},$$

for $x \in \mathbb{R}^k$. The map given by g will be one-to-one and onto if and only if the matrix g is invertible. Such matrices make up the *general linear group*

$$GL(k, \mathbb{R}) = \{\text{invertible } k \times k \text{ matrices}\}.$$

17.2.2. Definition of a Vector Bundle

We will start with the definition for a vector bundle. We want to capture the idea of attaching at each point of a manifold a vector space. We follow chapter 1, section 2 of Wells's *Differential Analysis on Complex Manifolds* [68].

Definition 17.2.1. *A real topological vector bundle of rank k over a manifold M is a manifold E and a continuous onto map*

$$\pi : E \to M$$

such that

1. *For each point $x \in M$ the inverse image of π is a rank k real vector space (i.e., $\pi^{-1}(x)$ is a rank k vector space).*
2. *For each $x \in M$, there is an open neighborhood U of x such that there is a homeomorphism*

$$h : \pi^{-1}U \to U \times \mathbb{R}^k$$

such that h restricted to the inverse image of x is a vector space isomorphism onto $\mathbb{R}^k$.

A vector bundle is an example of a *fiber bundle*. The inverse image over a point $x \in M$ is called the *fiber* and is denoted by E_x. By condition (1), the fiber E_x is a real vector space of dimension k. The underlying manifold M is called the *base space*.

We want to calculate on vector bundles and thus need to set up a language for local coordinates.

Here is another way of thinking of vector bundles. Start with our base manifold M. Cover M by a collection of open sets U_α. On each open set U_α, consider the product space

$$U_\alpha \times \mathbb{R}^k.$$

To form our vector bundle E, we want to glue, or patch, together the various $U_\alpha \times \mathbb{R}^k$ on any intersection $U_\alpha \cap U_\beta$. To fix notation, for each α, label the map $h : \pi^{-1}U_\alpha \to U_\alpha \times \mathbb{R}^k$ as h_α and the inverse map as

$$h_\alpha^{-1} : U_\alpha \times \mathbb{R}^k \to \pi^{-1}U_\alpha.$$

On the intersection of $U_\alpha \cap U_\beta$, define

$$g_{\alpha\beta} = h_\beta \circ h_\alpha^{-1}.$$

Then, on the intersection, we have

$$g_{\alpha\beta} = \text{Identity on } U_\alpha \cap U_\beta \times (\text{invertible } k \times k \text{ matrix})$$
$$= \text{Identity on } U_\alpha \cap U_\beta \times (\text{element of } GL(k, \mathbb{R})).$$

The $g_{\alpha\beta}$ are called *transition functions*. Often we will identify each transition function $g_{\alpha\beta}$ with its matrix. One can show, though we will not, the following:

Theorem 17.2.1. *For a vector bundle E, the transition functions satisfy*

1. *For all $(x, v) \in U_\alpha \times \mathbb{R}^k$,*

$$g_{\alpha\alpha}(x, v) = (x, v).$$

2. *For all $x \in U_\alpha \cap U_\beta \cap U_\gamma$, we have*

$$g_{\alpha\beta} \cdot g_{\beta\gamma} = g_{\alpha\gamma}.$$

(This last condition is known as the co-cycle condition; this and its analogs come up a lot in mathematics.)

Suppose we have a vector bundle E. On an open set U of the base manifold M, suppose we have a map

$$s : M \to E$$

such that

$$\pi \circ s = \text{Identity on } M.$$

The map s is called a *section* of E. On an open set U in M, we have a *framing* of E if we can find sections $s_1, \ldots, s_k$ such that for all $x \in U$, the vectors $s_1(x), \ldots, s_k(x)$ are linearly independent. Then the map

$$h^{-1} : U \times \mathbb{R}^k \to \pi^{-1}U$$

can be given by

$$h^{-1}(x, (a_1, \ldots, a_k)) = \sum a_i s_i(x).$$

Suppose we have an open set U_α with a framing $s_1^\alpha, \ldots, s_k^\alpha$ with corresponding map

$$h_\alpha^{-1} : U_\alpha \times \mathbb{R}^k \to \pi^{-1}U_\alpha$$

and have an open set U_β with a framing $s_1^\beta, \ldots, s_k^\beta$ such that for all $x \in U_\beta$ with corresponding map

$$h_\beta^{-1} : U_\alpha \times \mathbb{R}^k \to \pi^{-1}U_\beta.$$

We want to find the invertible matrix making up the transition function

$$g_{\alpha\beta} = h_\beta \circ h_\alpha^{-1} : U_\alpha \cap U_\beta \times \mathbb{R}^k \to U_\alpha \cap U_\beta \times \mathbb{R}^k.$$

(Keeping track of the various h_α versus the inverse map h_α^{-1} can be annoying; it is easy to get the various maps mixed up when doing a calculation.) At each point x in the intersection $U_\alpha \cap U_\beta$, we know that each $s_j^\alpha(x)$, as an element of the vector space E_x, can be written as a linear combination of the basis vectors $s_1^\beta(x), \ldots, s_k^\beta$. Thus there are numbers a_{ij} (depending on the point x) such that

$$s_j^\alpha(x) = \sum_{i=1}^k a_{ij} s_i^\beta(x).$$

Consider the matrix

$$g_{\alpha\beta} = \begin{pmatrix} a_{11} & \cdots & a_{1k} \\ \vdots & \vdots & \vdots \\ a_{k1} & \cdots & a_{kk} \end{pmatrix}.$$

Then if

$$g_{\alpha\beta}(x,(a_1,\ldots,a_k)) = h_\beta \circ h_\alpha^{-1}(x,(a_1,\ldots,a_k)) = (x,(b_1,\ldots,b_k)),$$

we have

$$\begin{pmatrix} b_1 \\ \vdots \\ b_k \end{pmatrix} = g_{\alpha\beta} \begin{pmatrix} a_1 \\ \vdots \\ a_k \end{pmatrix} = \begin{pmatrix} a_{11} & \cdots & a_{1k} \\ \vdots & \vdots & \vdots \\ a_{k1} & \cdots & a_{kk} \end{pmatrix} \begin{pmatrix} a_1 \\ \vdots \\ a_k \end{pmatrix}.$$

In working with vector bundles, there is the constant tension of writing everything out in terms of transition functions (which you must do frequently if you want to calculate anything) and thinking of the vector bundles as some sort of abstract manifold E. The reality is that one must be comfortable doing both.

17.3. Principal Bundles

One of the difficulties in this subject is that people use markedly different languages to describe what is in essence the same thing. In the last section we defined the notion of a vector bundle. Another approach is to use *principal bundles*, the topic of this section. We will first motivate the development of principal bundles from vector bundles and only then give the technical definition. We could easily have reversed the path, first defining principal bundles and then using this to motivate vector bundles.

Suppose we have a vector bundle E with base manifold M. Then for each point $p \in M$, there is the vector space E_p. In studying any type of mathematical object, one can either concentrate on the objects themselves or concentrate on the maps between these objects. In linear algebra, the objects are vector spaces while the maps are linear transformations. For vector bundles, this suggests that instead of considering the fiber of a point to be the vector space E_p, possibly we should have the fiber be the linear one-to-one, onto transformations from the vector space E_p to itself. The collection of all linear one-to-one, onto transformations from a vector space to itself forms a group called the *automorphism* group. Thus the fiber would be a group, not a vector space.

First, the groups will be Lie groups. A Lie group is simply a group that is also a differentiable manifold, with the requirement that the function given by the group operation must be smooth (i.e., for each fixed $g \in G$, the function that sends each $h \in G$ to gh must be smooth) and the function of taking inverses (taking each $g \in G$ to g^{-1}) must be smooth. For example, the group $GL(n, \mathbb{R})$ of invertible matrices is a Lie group.

Definition 17.3.1. *For a Lie group G, a* principal *G*-bundle *over a manifold M is a manifold P with a continuous onto map*

$$\pi : P \to M$$

such that

1. *For each point $p \in M$ the inverse image of π is a copy of the group G.*
2. *For each $x \in M$, there is an open neighborhood U of x such that there is a homeomorphism*

$$h : \pi^{-1}U \to U \times G$$

   such that h restricted to the inverse image of x is a group isomorphism onto G.

In the previous section, from the definition for vector bundles we constructed the transition functions $g_{\alpha\beta}$, which were elements of $GL(n, \mathbb{R})$ and hence can be used to construct a principal bundle. We could have reversed the direction and, starting with the transition functions $g_{\alpha\beta}$ and principal bundles, constructed vector bundles. In this book, we will overwhelmingly take the vector bundle approach. This is just a matter of choice.

17.4. Cylinders and Möbius Strips

We look at two examples of rank one vector bundles. Start with the cylinder

$$E = \{(x, y, z) \in \mathbb{R}^3 : x^2 + y^2 = 1\}.$$

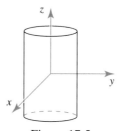

Figure 17.5

Given any (x, y) with $x^2 + y^2 = 1$, we can choose any number z and be guaranteed that (x, y, z) is on the cylinder. We can think of the base manifold M as the circle S^1 where $z = 0$:

Base Manifold $M = \{(x, y) : x^2 + y^2 = 1\} = \{(\cos(\theta), \sin(\theta)) : \theta \in \mathbb{R}\}$,

which we can identify with

$$\{(x, y, 0) : x^2 + y^2 = 1\} = \{(\cos(\theta), \sin(\theta), 0) : \theta \in \mathbb{R}\}.$$

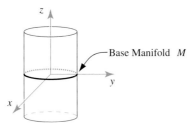

Figure 17.6

The fiber over any point $p = ((\cos(\theta), \sin(\theta)) \in M$ is the entire line parallel to the z-axis.

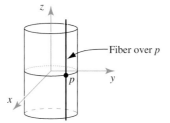

Figure 17.7

The Möbius strip looks like

Figure 17.8

The base manifold M will again be the circle, which we can identify to

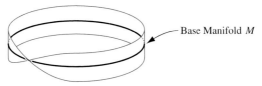

Figure 17.9

Consider the circle with the open sets U_1, U_2 and U_3.

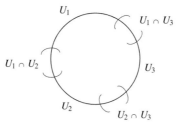

Figure 17.10

We will define the Möbius strip via transition functions g_{12}, g_{13} and g_{23}. We need

$$g_{\alpha\beta} : U_\alpha \cap U_\beta \times \mathbb{R} \to U_\alpha \cap U_\beta \times \mathbb{R}.$$

Set

$$g_{12}(p,v) = (p,v)$$
$$g_{13}(p,v) = (p,v)$$
$$g_{23}(p,v) = (p,-v).$$

Visually, we have Figure 17.11.

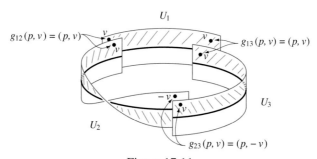

Figure 17.11

Note that if the last map were $g_{23}(p,v) = (p,v)$, we would have ended up with the cylinder.

17.5. Tangent Bundles

17.5.1. Intuitions

One of the most important vector bundles is the *tangent bundle* of a manifold, which is the subject of the rest of this chapter.

Intuitively, the tangent bundle $T(M)$ of a manifold M should be the space made up of all tangent spaces for M at each point of M. For example, consider the circle S^1 in the plane. We want the fiber $T_p(S^1)$ of any point $p \in S^1$ to be the tangent line, thought of as the real line with origin at p. Thus $T_{(1,0)}(S^1)$ is a copy of the real line parallel to the y-axis. Similarly, $T_{(0,1)}(S^1)$ and $T_{(1/2,\sqrt{3}/2)}(S^1)$ are (Figure 17.12):

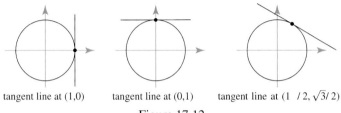

tangent line at (1,0)   tangent line at (0,1)   tangent line at (1 / 2, $\sqrt{3}$/ 2)

Figure 17.12

In a similar fashion, we want the tangent bundle $T(S^2)$ for the unit sphere S^2 in $\mathbb{R}^3$ to be all the tangent planes of points of S^2. Thus $T_{(1,0,0)}(S^2)$ is (Figure 17.13)

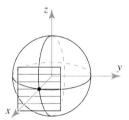

Figure 17.13

while $T_{(1/\sqrt{3},1/\sqrt{3},1/\sqrt{3})}(S^2)$ is (Figure 17.14)

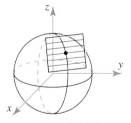

Figure 17.14

For manifolds defined in some ambient $\mathbb{R}^n$, all of this can be done in a fairly straightforward way, as we will do in the next section for parametrically defined manifolds.

But what if we have an abstract manifold, one that does not "live" in some $\mathbb{R}^n$? Here, the answer is to think of tangent spaces not so much as geometric objects as more like derivatives. Recall that one of the main goals of calculus is to find the slope of the tangent line to a curve $y = f(x)$.

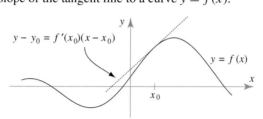

$y - y_0 = f'(x_0)(x - x_0)$

Figure 17.15

Of course, the slope is just the derivative $f'(x)$, giving us that the line tangent to the curve at a point (x_0, y_0) is

$$y - y_0 = f'(x_0)(x - x_0).$$

Similarly, for a surface in $\mathbb{R}^3$ given as a zero set of a function $f(x, y, z) = 0$, we know from multivariable calculus that a normal vector at a point $p = (x_0, y_0, z_0)$ on the surface is

$$\nabla f(p) = \left(\frac{\partial f}{\partial x}(p), \frac{\partial f}{\partial y}(p), \frac{\partial f}{\partial z}(p) \right),$$

giving us that the tangent plane at the point is

$$\frac{\partial f}{\partial x}(p)(x - x_0) + \frac{\partial f}{\partial y}(p)(y - y_0) + \frac{\partial f}{\partial z}(p)(z - z_0) = 0.$$

This suggests a more intrinsic way for defining the tangent bundle, using some notion of derivative. This we do in the last section of this chapter.

17.5.2. Tangent Bundles for Parametrically Defined Manifolds

This definition of tangent bundle will not be that hard, as we have basically set up the correct notation for the tangent bundle in Section 16.2.1. As before, let $x = (x_1, x_2, \ldots, x_n)$ be the coordinates for $\mathbb{R}^n$ and let $u = (u_1, \ldots, u_k)$ be coordinates for $\mathbb{R}^k$. If M is a k-dimensional manifold living in an ambient $\mathbb{R}^n$, then we know that for every point $p \in M$ there are an open ball V in $\mathbb{R}^n$, centered at p, and a differentiable map

$$\gamma : U \rightarrow \mathbb{R}^n,$$

with U an open ball in $\mathbb{R}^k$, centered at the origin, such that

1. $\gamma(0,\ldots,0) = p$.
2. γ is a one-to-one onto map from U to $M \cap V$.
3. The rank of the Jacobian for γ is k at every point.

Each row of the Jacobian is a vector in $\mathbb{R}^n$. Further, by the rank condition, we know that the k rows are linearly independent. The tangent space $T_p M$ is the k-dimensional vector space spanned by the rows of the Jacobian. (All of this is done in multivariable calculus, though the usual examples are of curves in the plane, curves in space, or surfaces in space.)

Let us look at the circle again. Unlike in Chapter 16, consider the parameterization

$$\gamma : \mathbb{R} \to S^1$$

given by

$$\gamma(\theta) = (\cos(\theta), \sin(\theta)).$$

Then the Jacobian is

$$J(\gamma) = \left(\frac{\partial x_1}{\partial \theta}, \frac{\partial x_2}{\partial \theta} \right) = (-\sin(\theta), \cos(\theta)).$$

Note that at $p = (1,0)$, or when $\theta = 0$, the tangent line is spanned by

$$(-\sin(0), \cos(0)) = (0,1),$$

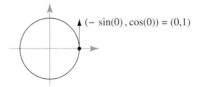

Figure 17.16

as we want.

A given parameterization for a manifold need not be unique. Thus we should now show that the vector spaces spanned by the Jacobians for two different parameterizations agree at any given point. The proof, which we will not do, is simply an application of the chain rule. This should be no surprise, as different parameterizations can be thought of as different coordinate systems on the manifold.

17.5.3. $T(\mathbb{R}^2)$ as Partial Derivatives

Let us start with the plane $\mathbb{R}^2$. At any point $(x_0, y_0) \in \mathbb{R}^2$, the tangent plane $T_{(x_0, y_0)} \mathbb{R}^2$ is just another copy of $\mathbb{R}^2$.

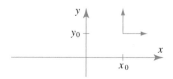

Figure 17.17

But we now want to emphasize tangency as derivatives, or as rates of change. The partial derivative $\partial/\partial x$ acting on any function $f(x,y)$, evaluated at (x_0, y_0), measures how fast f is changing at (x_0, y_0) in the $(1,0)$ direction,

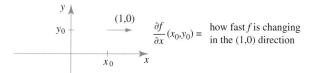

Figure 17.18

while $\partial/\partial y$ acting on $f(x,y)$, evaluated at (x_0, y_0), measures how fast f is changing at (x_0, y_0) in the $(0,1)$ direction. In general, Figure 17.19

$$\left(a\frac{\partial}{\partial x} + b\frac{\partial}{\partial y} \right) f \Big|_{(x_0, y_0)}$$

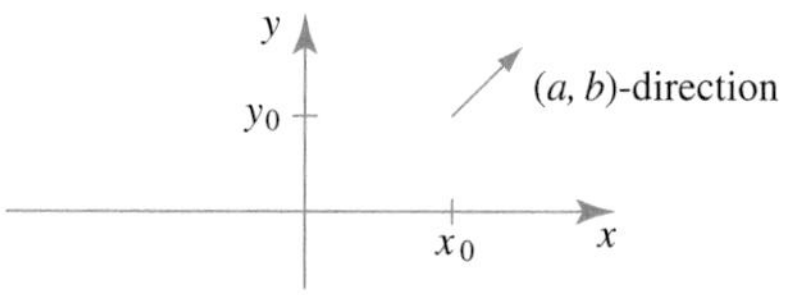

Figure 17.19

gives the rate of change of f at (x_0, y_0) in the (a,b) direction. We will identify the tangent plane of $(x_0, y_0) \in \mathbb{R}^2$ as the two-dimensional vector space

$$T_{(x_0,y_0)}\mathbb{R}^2 = \left\{ a\frac{\partial}{\partial x} + b\frac{\partial}{\partial y} \Big|_{(x_0,y_0)} : a, b \in \mathbb{R} \right\}.$$

As will be shown in the exercises,

Theorem 17.5.1. *For any $L \in T_{(x_0,y_0)}\mathbb{R}^2$, for all functions $f(x,y)$ and $g(x,y)$ and all real constants α and β, we have*

$$L(\alpha f + \beta g) = \alpha L(f) + \beta L(g)$$

$$L(fg) = fL(g) + gL(f).$$

(This actually follows from some simple properties of differentiation.) The key for us is that we can identify the tangent plane at a point in $\mathbb{R}^2$ with linear

maps
$$L : \text{Functions on } \mathbb{R}^2 \to \mathbb{R}$$

that satisfy Leibniz's rule $L(fg) = fL(g) + gL(f)$.

17.5.4. Tangent Space at a Point of an Abstract Manifold

Let M be an abstract n-dimensional manifold. We want to develop the idea that the elements of the tangent space at a point p will be ways for measuring the rates of change of a function on M.

We know that M has a cover $\mathcal{U} = \{U_\alpha\}$ with corresponding collection $\mathcal{V} = \{V_\alpha\}$ of connected open subsets of $\mathbb{R}^n$ and one-to-one onto maps $\chi_\alpha : V_\alpha \to U_\alpha$, such that, for any two open U_α and U_β, the map

$$\chi_\beta^{-1} \circ \chi_\alpha : \chi_\alpha^{-1}(U_\alpha \cap U_\beta) \to \chi_\beta^{-1}(U_\alpha \cap U_\beta)$$

is differentiable. Recall that this allowed us to define a real-valued function f on a manifold M to be differentiable if the composition function

$$f \circ \phi_\alpha : \text{Open ball in } \mathbb{R}^n \to \mathbb{R}$$

is differentiable as a real-valued function defined on $\mathbb{R}^n$.

We want to define what it means to differentiate functions on M. (The rest of this paragraph is a slight rephrasing of section 6.5.2 in [26].) Thus we want to measure the rate of change of f at p. This should only involve the values of f near p. The values f away from p should be irrelevant. This is the motivation behind the following equivalence relation. Let (f_1, U_1) and (f_2, U_2) denote open sets on M containing p, with f_1 and f_2 corresponding differentiable functions. We will say that

$$(f_1, U_1) \sim (f_2, U_2)$$

if, on the open set $U_1 \cap U_2$, we have $f_1 = f_2$. This leads us to defining

$$C_p^\infty = \{(f, U)\} / \sim.$$

We will frequently abuse notation and denote an element of C_p^∞ by f. The space C_p^∞ is a vector space and captures the properties of functions close to the point p. (For mathematical culture's sake, C_p^∞ is an example of a germ of a sheaf, in this case, the sheaf of differentiable functions.)

Definition 17.5.1. *The tangent space $T_p(M)$ is the space of all linear maps*

$$v : C_p^\infty \to C_p^\infty$$

such that

$$v(fg) = fv(g) + gv(f).$$

We say that the rate of change of a function f at a point $p \in M$ in the "direction" $v \in T_p(M)$ is the value of the function $v(f)$ at p.

Showing that $T_p(M)$ is a dimension n vector space is one of the exercises. Finding a relatively straightforward method for actually constructing elements of $T_p(M)$ is one of the goals of the next section.

17.5.5. Tangent Bundles for Abstract Manifolds

Intuitively, an abstract n-dimensional manifold M can be covered by various open sets U such that for each U there are an open ball $V \subset \mathbb{R}^n$ and a one-to-one onto map

$$\chi : V \to U.$$

Let $x_1, \ldots, x_n$ be local coordinates for $\mathbb{R}^n$. On V there are the partial derivatives $\partial/\partial x_1, \ldots, \partial/\partial x_n$. People like to think of U as being identified with the ball V, in which case on U there should also be corresponding partial derivatives. We want to write down here the explicit maps that are needed to make this rigorous.

We start with a function

$$f : U \to \mathbb{R}$$

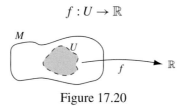

Figure 17.20

Corresponding to each partial derivative $\partial/\partial x_i$, we want to define an element $v_i \in T_p(M)$. Using our map $\chi : V \to U$, we know that

$$f \circ \chi : V \to \mathbb{R}.$$

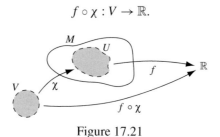

Figure 17.21

Then taking the partial derivative of $f \circ \chi$ with respect to the variable x_i makes perfect sense, giving us a new function

$$\frac{\partial (f \circ \chi)}{\partial x_i} : V \to \mathbb{R}.$$

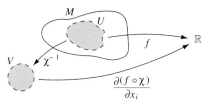

Figure 17.22

Then we define

$$v_i(f) = \frac{\partial (f \circ \chi)}{\partial x_i} \circ \chi^{-1} : U \to \mathbb{R}.$$

Showing that this v_i is indeed in $T_p(M)$ is left to the exercises, as is showing that the v_i are linearly independent. Not left to the exercises, but still passed over, is the proof that the various v_i form a basis for $T_p(M)$.

This notation is admittedly a bit complicated. Since χ is a one-to-one onto map, people frequently just identify the open set U with the open set V in $\mathbb{R}^n$, and let $x_1, \ldots, x_n$ denote the coordinates on U, without putting in the required maps χ and χ^{-1}. When this is done, people consider the partial derivatives

$$\frac{\partial}{\partial x_i}$$

as forming the basis of the tangent bundle for M over the open set U. In the next chapter, for example, we will write a tangent vector on the open set U of a manifold M as some

$$a_1 \frac{\partial}{\partial x_1} + \cdots + a_n \frac{\partial}{\partial x_n}.$$

Finally, the coordinate maps $\chi : U \to V$ are hardly unique. Showing that we get the same vector space $T_p(M)$ for different coordinate maps is a non-trivial application of the chain rule.

17.6. Exercises

Exercise 17.6.1. *For the unit circle S^1 in $\mathbb{R}^2$, consider the following three parameterizations:*

$$\gamma_1(\theta) = (\cos(\theta), \sin(\theta))$$
$$\gamma_2(s) = (s, \sqrt{1-s^2})$$
$$\gamma_3(t) = (\sqrt{1-t^2}, t).$$

Show that

$$\gamma_1(\pi/4) = \gamma_2(1/\sqrt{2}) = \gamma_3(1/\sqrt{2}).$$

At this common point on S^1, compute the Jacobian for each of these parameterizations. Show that at this point, each Jacobian spans the same vector space.

Exercise 17.6.2. *For the unit sphere S^2 in $\mathbb{R}^3$, consider the parameterization*

$$\gamma(u,v) = (u, v, \sqrt{1-u^2-v^2})$$

from the unit disc $\{(u,v) : u^2 + v^2 < 1\}$ to the hemisphere. Compute the Jacobian.

Exercise 17.6.3. *Using the notation from the previous exercise, show that the tangent space of S^2 at $(0,0,1)$ is a plane parallel to the xy-plane.*

Exercise 17.6.4. *Using the previous notation, find the Jacobian at the point $\gamma(1/\sqrt{3}, 1/\sqrt{3})$.*

Exercise 17.6.5. *Consider a different parameterization for S^2:*

$$\tau(s,t) = (s, \sqrt{1-s^2-t^2}, t).$$

Find the Jacobian at $\tau(1/\sqrt{3}, 1/\sqrt{3})$. Show that the rows of this Jacobian (which form a basis for the tangent plane at this point) span the same vector space as the rows of the Jacobian in the previous problem.

Exercise 17.6.6. *Let*

$$T_{(x_0,y_0)}\mathbb{R}^2 = \left\{ a\frac{\partial}{\partial x} + b\frac{\partial}{\partial y}\Big|_{(x_0,y_0)} : a, b \in \mathbb{R} \right\}.$$

Show for any $L \in T_{(x_0,y_0)}\mathbb{R}^2$ that for all functions $f(x,y)$ and $g(x,y)$ and all real constants α and β we have

$$L(\alpha f + \beta g) = \alpha L(f) + \beta L(g)$$
$$L(fg) = fL(g) + gL(f).$$

Exercise 17.6.7. *Show that $T_p(M)$, the space of all linear maps*

$$v : C_p^\infty \to C_p^\infty$$

such that

$$v(fg) = f v(g) + g v(f),$$

is a vector space.

Exercise 17.6.8. *If*

$$v_i(f) = \frac{\partial (f \circ \chi)}{\partial x_i} \circ \chi^{-1} : U \to \mathbb{R},$$

show that $v_i \in T_p(M)$.

Exercise 17.6.9. *Using the notation from the previous exercise, show that the $v_1, \ldots, v_n$ are linearly independent.*

18

Connections

Summary: The goal of this chapter is to develop the technical definition for a connection on a vector bundle, which will allow us to differentiate sections of a vector bundle. Physicists usually refer to connections as "gauges." As we will see in later chapters, choosing scalar and vector potentials can be recast as choosing a connection for an appropriate vector bundle.

18.1. Intuitions

Suppose we have a vector bundle E over a base manifold M. Let

$$s : M \to E$$

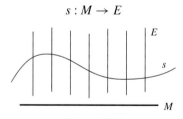

Figure 18.1

be a section. We want to determine how fast the section is varying in E. We would like to take the derivative of s. Surprisingly, this is not possible. Naively, the derivative at a point p would be

$$\lim_{q \to p} \frac{s(q) - s(p)}{q - p}.$$

Problems immediately arise with making sense out of this formula. First, p and q are points in a manifold; $q - p$ does not make sense. This is not the main problem, as we could instead try to measure the rate of change of s along a parameterized path $\sigma(t)$ in M with $\sigma(0) = p$ and then correspondingly alter our inital stab at differentiating s by trying to calculate

$$\lim_{t \to 0} \frac{s(\sigma(t)) - s(p)}{t}.$$

We must make sense out of the numerator $s(\sigma(t)) - s(p)$. This is the real difficulty, as $s(p)$ is in the vector space E_p and $s(\sigma(t))$ is in the vector space $E_{\sigma(t)}$. These are obviously different vector spaces. If we want to measure how fast our section is changing, then we must somehow measure changes in different vector spaces. We must somehow link up or "connect" the different spaces. This will lead to the notion of *connection*.

Before giving the definition, let us emphasize the problem. In a vector bundle E over a base space M of rank k, it is certainly the case that the fiber over any point in M is a real k-dimensional vector space. Thus the vector spaces E_p and E_q are isomorphic. But this does not allow us to compare vectors in E_p with vectors in E_q, since these vector spaces are not canonically isomorphic (meaning there is no intrinsic one-to-one onto linear transformation between them). There is no intrinsic way, independent of arbitrary choices, for comparing the vectors. There are many different possible isomorphisms between the vector spaces E_p and E_q. None are intrinsically better than any others.

Thus it is not surprising that there will be no single method for measuring the rate of change of a section. It is also the case that there is more than one method for defining connections (though all are equivalent). We will characterize, in an algebraic fashion, what these rates of change can be, in a way that is analogous to defining the derivative not in terms of limits, but instead as a linear map from function spaces to function spaces that maps constants to zero and satisfy Leibniz's rule. Then we will return to a more limit type approach, where we develop the notion of parallel transport of a vector along a curve in the base manifold.

18.2. Technical Definitions

18.2.1. *Operator Approach*

(We will be using tensor products in this section; these are defined in the Appendix of this chapter.)

For a vector bundle E with base space M, let

$$\Gamma(E) = \{s : M \to E : \pi \circ s = \text{identity map}\}$$

denote the space of all differentiable sections from M to E, where $\pi : E \to M$ is the map that sends every point in the fiber E_p to the point p in M. Also, recall that $\Lambda^k(M)$ denotes the k-forms on M.

Definition 18.2.1. *A connection on a vector bundle E with base space M is any linear map*

$$\nabla : \Lambda^k(M) \otimes \Gamma(E) \to \Lambda^{k+1}(M) \otimes \Gamma(E)$$

that satisfies, for all k forms ω on M and sections $s \in \Gamma(E)$,

$$\nabla(\omega \otimes s) = \omega \wedge \nabla(s) + (-1)^k d\omega \otimes s.$$

We will concentrate on the case

$$\nabla : \Gamma(E) \to \Lambda^1(M) \otimes \Gamma(E).$$

(Recall that $\Lambda^0(M)$ are constants, so that $\Lambda^0(M) \otimes \Gamma(E) = \Gamma(E)$.) Let $f : M \to \mathbb{R}$ be any differentiable function on M. Thus for each point $p \in M$, $f(p)$ is a number. Now let $s \in \Gamma(E)$ be a section of E. Then $s(p)$ is a vector in the vector space E_p. We can multiply this vector by any scalar, and in particular multiply the vector $s(p)$ by the number $f(p)$. Hence in a natural way, fs is another section of the bundle E, with

$$(fs)(p) = f(p)s(p) \in E_p.$$

A connection must have the property that

$$\nabla(fs) = f\nabla(s) + (df) \otimes s,$$

which is just a version of the single variable calculus requirement that the derivative of a product satisfy

$$(uv)' = u'v + v'u \text{ (Leibniz's rule).}$$

We want to understand how to create and how to write down connections. Over an open set U in the base manifold M, let $s_1, \ldots, s_k$ be a framing for the vector bundle E. (Thus at each point $p \in M$, the vectors $s_1(p), \ldots, s_k(p)$ must form a basis of the vector space E_p.) Let ∇ be a connection. Then for each section s_i, we know that ∇s_i must be an element of $\Lambda^1(M) \otimes \Gamma(E)$. Thus there must be 1-forms ω_{ij} on U such that

$$\nabla s_i = \sum_{j=1}^{k} \omega_{ij} \otimes s_j.$$

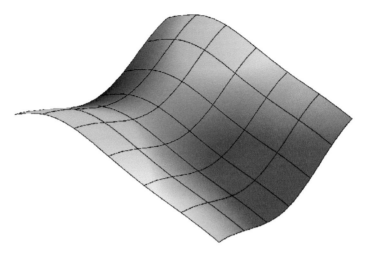

Plate 1

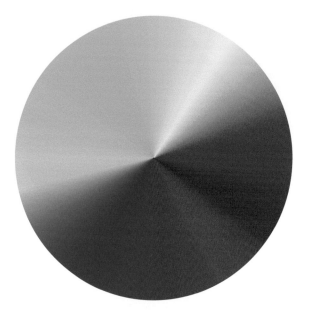

Plate 2

Our connection may thus be described as a $k \times k$ matrix of 1-forms

$$\omega = (\omega_{ij}).$$

This matrix ω depends on our choice of a framing (or of a basis) for the vector bundle E.

We can describe $\nabla(s)$ for any section $s \in \Gamma(E)$ in terms of the connection matrix ω as follows. Since the $s_1, \ldots, s_k$ form a basis of sections, there are real-valued functions f_i defined on M such that

$$s = f_1 s_1 + \cdots + f_k s_k.$$

Then we have

$$\begin{aligned}
\nabla(s) &= \nabla(f_1 s_1 + \cdots + f_k s_k) \\
&= \nabla(f_1 s_1) + \cdots + \nabla(f_k s_k) \\
&= f_1 \nabla(s_1) + s_1 \mathrm{d}(f_1) + \cdots + f_k \nabla(s_k) + s_k \mathrm{d}(f_k) \\
&= f_1 \left(\sum_{j=1}^{k} \omega_{ij} \otimes s_j \right) + s_1 \mathrm{d}(f_1) + \cdots + f_k \left(\sum_{j=1}^{k} \omega_{kj} \otimes s_j \right) + s_k \mathrm{d}(f_k) \\
&= \left(\sum_{i=1}^{k} f_i \omega_{i1} + \mathrm{d} f_1 \right) s_1 + \cdots + \left(\sum_{i=1}^{k} f_i \omega_{ik} + \mathrm{d} f_k \right) s_k.
\end{aligned}$$

(Note that in the last equation, for notational convenience, we are suppressing the $\otimes$.)

As can be seen, such direct calculations are not pleasant. Here is a more user-friendly style of notation, though it must be emphasized that actually to calculate, one must usually write everything out as we did in the previous example.

To ease notation, suppose that E is a rank-two vector bundle. Let s_1 and s_2 be two linearly independent sections. Thus every section S can be written as

$$S = f_1 s_1 + f_2 s_2,$$

where f_1 and f_2 are real-valued functions on the base manifold M. Set

$$f = \begin{pmatrix} f_1 & f_2 \end{pmatrix}, \ s = \begin{pmatrix} s_1 \\ s_2 \end{pmatrix}.$$

Note that we are now using the symbol s to denote a column vector. That is why we use a capital S to denote the section. We have

$$S = f \cdot s.$$

Then the connection can be written as

$$\nabla(S) = \nabla(f \cdot s)$$
$$= df \cdot s + f \cdot \nabla s$$
$$= df \cdot s + f \cdot \omega \cdot s,$$

which is just a slightly slicker way of writing

$$\nabla(f_1 s_1 + f_2 s_2) = \begin{pmatrix} df_1 & df_2 \end{pmatrix} \begin{pmatrix} s_1 \\ s_2 \end{pmatrix} + \begin{pmatrix} f_1 & f_2 \end{pmatrix} \begin{pmatrix} \omega_{11} & \omega_{12} \\ \omega_{21} & \omega_{22} \end{pmatrix} \begin{pmatrix} s_1 \\ s_2 \end{pmatrix}.$$

In general, a section S of a rank k bundle can be written as the product $S = f \cdot s$, where f denotes a $1 \times k$ row matrix of functions and s denotes a $k \times 1$ column vector.

All of the preceding calculations depend on choosing a particular basis of sections. If we choose a different basis, then the connection matrix will be different, despite the fact that both are representing the same underlying connection. We now want to see how to compare different connection matrices with respect to different choices of frames.

Let $s_1, \ldots, s_k$ be a framing with connection matrix ω. Let $t_1, \ldots, t_k$ be a different framing, with connection matrix τ. Let A be the $k \times k$ invertible matrix of functions that takes the t frame to the s frame, namely,

$$s = At$$

or, equivalently,

$$s_1 = a_{11} t_1 + \cdots + a_{1k} t_k$$

$$\vdots$$

$$s_k = a_{k1} t_1 + \cdots + a_{kk} t_k.$$

Our goal is to show

Proposition 18.2.1.

$$\omega A = A\tau + dA,$$

which is equivalent to showing for each i and j that

$$\omega_{i1} a_{1j} + \cdots + \omega_{ik} a_{kj}$$

is equal to

$$a_{i1} \tau_{ij} + \cdots + a_{ik} \tau_{kj} + da_{ij}.$$

Proof. Let S be a section. Then we can write S either as a linear combination of the s_i or as a linear combination of the t_i. We write

$$S = f \cdot s = g \cdot t.$$

Here f and g are row vectors of functions. Since $s = At$, we have

$$f \cdot s = f \cdot A \cdot t = g \cdot t,$$

which means that

$$g = f \cdot A.$$

Now

$$
\begin{aligned}
\nabla(S) &= \nabla(f \cdot s) \\
&= \mathrm{d}f \cdot s + f \cdot \omega \cdot s \\
&= \mathrm{d}f \cdot A \cdot t + f \cdot \omega \cdot A \cdot t \\
&= (\mathrm{d}f \cdot A + f \cdot \omega \cdot A) \cdot t.
\end{aligned}
$$

But we also have

$$
\begin{aligned}
\nabla(S) &= \nabla(g \cdot t) \\
&= \mathrm{d}g \cdot t + f \cdot \tau \cdot t \\
&= (\mathrm{d}g + g \cdot \tau) \cdot t \\
&= (\mathrm{d}(f \cdot A) + f \cdot A \cdot \tau) \cdot t \\
&= (\mathrm{d}f \cdot A + f \cdot \mathrm{d}A + f \cdot A \cdot \tau) \cdot t.
\end{aligned}
$$

Thus we have

$$f \cdot \mathrm{d}A + f \cdot A\tau = f \cdot \omega \cdot A,$$

which means that $\omega A = A\tau + \mathrm{d}A$.

$\square$

We could have written the preceding out in local coordinates and just computed. This would have the advantage of being a bit more concrete but the disadvantage of looking messy.

18.2.2. Connections for Trivial Bundles

In the last thirty or so years, the study of the space of possible connections for a vector bundle has become increasingly central to both modern mathematics and physics. This is not a subject to be entered into lightly. In this section we will look at the simplest class of examples, namely, connections on trivial bundles. Though we are using the term "trivial," these types of connections will lead to a natural bundle interpretation of Maxwell's equations in Chapter 20.

Definition 18.2.2. *A rank* k *vector bundle* E *over a base manifold* M *is trivial if*

$$E = M \times \mathbb{R}^k.$$

Thus the cylinder from the last chapter is a trivial bundle. Though it is not particularly natural, we can interpret the plane $\mathbb{R}^2$ as a rank-one bundle over $\mathbb{R}$, treating the x-axis as the base manifold and the vertical lines as the fibers.

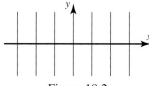

Figure 18.2

In a similar way, as you are asked to discuss in the exercises, we can think of $\mathbb{R}^3$ as a rank-two bundle over the line or as a rank-one bundle over the plane.

In terms of transition functions, we have

Theorem 18.2.1. *A vector bundle* E *over a base manifold* M *is trivial if and only if we can find an open covering* $\{U_\alpha\}$ *of* M *such that the transition functions are always the identity map, that is,*

$$g_{\alpha\beta} = Identity.$$

Though it is not particularly hard, we will not prove this.

For a trivial bundle $M \times \mathbb{R}^k$, the sections are quite easy to write down. All we need to do is to choose k real-valued functions

$$f_i : M \to \mathbb{R}.$$

The corresponding section

$$s : M \to M \times \mathbb{R}^k$$

is simply

$$s(x) = \left(x, \begin{pmatrix} f_1(x) \\ \vdots \\ f_k(x) \end{pmatrix}\right).$$

Connections are fairly easy to describe for trivial vector bundles. We only need to specify the $k \times k$ matrix ω of 1-forms.

Definition 18.2.3. *The* trivial connection *for a trivial vector bundle occurs when*

$$\omega = (0).$$

To be more specific, given a section $s : M \to M \times \mathbb{R}^k$, the trivial connection is

$$\nabla = d$$

and thus

$$\nabla s = ds = \begin{pmatrix} d f_1(x) \\ \vdots \\ d f_k(x) \end{pmatrix}.$$

To be even more specific, if it happens that the base manifold M is the real numbers $\mathbb{R}$ with coordinate x, then

$$\nabla s = ds = \begin{pmatrix} d f_1(x) \\ \vdots \\ d f_k(x) \end{pmatrix} = \begin{pmatrix} f_1'(x) dx \\ \vdots \\ f_k'(x) dx \end{pmatrix}.$$

There are connections on trivial bundles that are not the trivial connection. This will become important when we put Maxwell's equations into the language of vector bundles. For now, suppose $E = M \times \mathbb{R}^k$ is a trivial bundle. Then we can define a connection by choosing any $k \times k$ matrix of 1-forms on M and setting

$$\nabla = d + \omega.$$

For example, let E be the trivial rank-one bundle over the real line $\mathbb{R}$ with coordinate x. Choose our matrix of one-forms to be

$$\omega = (x^2 dx).$$

Since E has rank-one, a section is just a single real-valued function $f(x)$. Then

$$\nabla(f) = (d + \omega)(f(x)) = (f'(x) + x^2 f(x)) dx.$$

For another example, now let E be the trivial rank-two bundle over the real line $\mathbb{R}$ with coordinate x. Choose the connection 1-form matrix to be

$$\omega = \begin{pmatrix} x dx & e^x dx \\ dx & x^2 dx \end{pmatrix}.$$

A section is now given by two real-valued functions $f_1(x)$ and $f_2(x)$. We have

$$\nabla(s) = \nabla \begin{pmatrix} f_1(x) \\ f_2(x) \end{pmatrix}$$

$$= \left(d + \begin{pmatrix} x\,dx & e^x\,dx \\ dx & x^2\,dx \end{pmatrix} \right) \begin{pmatrix} f_1(x) \\ f_2(x) \end{pmatrix}$$

$$= \begin{pmatrix} (f_1' + xf_1 + e^x f_2)dx \\ (f_2' + f_1 + x^2 f_2)dx \end{pmatrix}.$$

18.3. Covariant Derivatives of Sections

(While all of this section is standard, we are following the ideas in [10].) Our original goal was to differentiate sections. Our claim is that the technical definition given in the last section of a connection as a linear map

$$\nabla : \Gamma(E) \to \Lambda^1(M) \otimes \Gamma(E)$$

is the correct definition for the derivative of a section.

Let $s \in \Gamma(E)$ be a section. Let

$$\sigma : [a,b] \to M$$

be a curve in the base manifold M. Our goal is to fix a connection ∇ and to use ∇ to differentiate the section s along the curve σ. First, to simplify matters, we will assume that the image of σ lands in an open set U of M that we can identify with an open ball in $\mathbb{R}^n$. This allows us to fix local coordinates $x_1, \ldots, x_n$ for U, which in turn allows us to describe the curve σ by n functions:

$$\sigma(t) = (x_1(t), \ldots, x_n(t)).$$

Then the tangent vector to our curve is

$$\frac{dx_1}{dt} \cdot \frac{\partial}{\partial x_1} + \cdots + \frac{dx_n}{dt} \cdot \frac{\partial}{\partial x_n}.$$

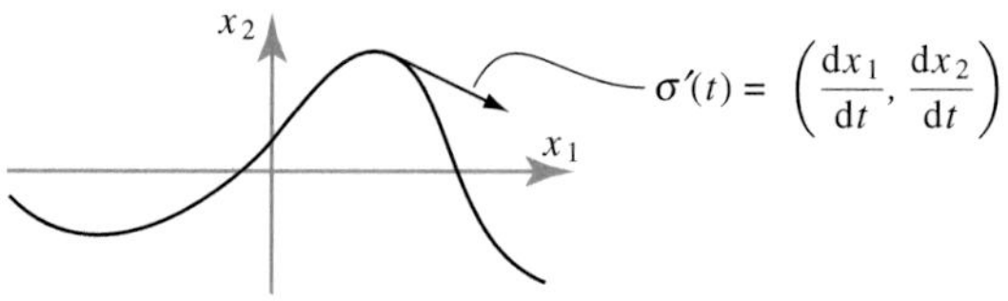

$$\sigma'(t) = \left(\frac{dx_1}{dt}, \frac{dx_2}{dt} \right)$$

Figure 18.3

Recall that 1-forms act on partial derivatives via the rule

$$dx_i \left(\frac{\partial}{\partial x_j} \right) = \begin{cases} 1 & \text{if } i = j \\ 0 & \text{if } i \neq j \end{cases}$$

and linearity. For example, consider the 1-form

$$\omega = x_1 dx_1 + (x_1 + x_2) dx_2.$$

At the point $(x_1, x_2) = (3, 2)$, we have

$$\omega = 3 dx_1 + 5 dx_2.$$

Let

$$v = x_2 \frac{\partial}{\partial x_1} + \frac{\partial}{\partial x_2},$$

which at the point $(3, 2)$ becomes

$$v = 2 \frac{\partial}{\partial x_1} + \frac{\partial}{\partial x_2}.$$

Then at the point $(3, 2)$, we have

$$\omega(v) = (3 dx_1 + 5 dx_2) \left(2 \frac{\partial}{\partial x_1} + \frac{\partial}{\partial x_2} \right) = 3 \cdot 2 + 5 \cdot 1 = 11.$$

The key is that 1-forms send tangent vectors to numbers.
   This suggests

Definition 18.3.1. *Let ∇ be a connection on a vector bundle E, let s be a section of E, and let X be a vector field on M. Then the* covariant derivative *of s with respect to X is*

$$\nabla_X(s) = \nabla(s)(X).$$

We will first make sense out of the preceding, and then see how actually to compute a covariant derivative. We know that $\nabla(s)$ is in $\Lambda^1(M) \otimes \Gamma(E)$ and thus consists of a linear combination of sections of E with 1-forms on M. These 1-forms act on the vector field X.
   If we want to differentiate along a curve σ, then we choose our vector field to be the tangent vectors to σ.
   Now to see how to compute. As we saw earlier, given a basis $s_1, \ldots, s_k$ for our vector bundle, for any section s there are real-valued functions f_i defined on M such that

$$s = f_1 s_1 + \cdots + f_k s_k.$$

We saw that

$$\nabla(s) = \left(\sum_{i=1}^{k} f_i \omega_{i1} + \mathrm{d} f_1 \right) s_1 + \cdots + \left(\sum_{i=1}^{k} f_i \omega_{ik} + \mathrm{d} f_k \right) s_k.$$

For any vector field X, we have

$$\nabla_X(s) = \left(\sum_{i=1}^{k} f_i \omega_{i1}(X) + \mathrm{d} f_1(X) \right) s_1 + \cdots + \left(\sum_{i=1}^{k} f_i \omega_{ik}(X) + \mathrm{d} f_k(X) \right) s_k.$$

(Just to be concrete, both $\sum_{i=1}^{k} f_i \omega_{i1}(X) + \mathrm{d} f_1(X)$ and $\sum_{i=1}^{k} f_i \omega_{ik}(X) + \mathrm{d} f_k(X)$ are numbers.)

We now look at a concrete example. Let E be a rank-two bundle with basis of section s_1 and s_2 on a surface M, which has coordinates x_1 and x_2. Let the connection matrix be

$$\omega = \begin{pmatrix} \mathrm{d} x_1 & \mathrm{d} x_1 + \mathrm{d} x_2 \\ \mathrm{d} x_2 & \mathrm{d} x_1 \end{pmatrix}.$$

Let the coefficients for our section s be

$$f_1(x_1, x_2) = x_1 + x_2$$
$$f_2(x_1, x_2) = x_1 - x_2.$$

Then

$$\mathrm{d} f_1 = \mathrm{d} x_1 + \mathrm{d} x_2$$
$$\mathrm{d} f_2 = \mathrm{d} x_1 - \mathrm{d} x_2.$$

Then

$$\nabla(s) = \nabla(f_1 s_1 + f_2 s_2)$$
$$= ((x_1 + x_2)\mathrm{d} x_1 + (x_1 - x_2)\mathrm{d} x_2 + (\mathrm{d} x_1 + \mathrm{d} x_2)) s_1$$
$$+ ((x_1 + x_2)(\mathrm{d} x_1 + \mathrm{d} x_2) + (x_1 - x_2)\mathrm{d} x_1 + (\mathrm{d} x_1 - \mathrm{d} x_2)) s_2.$$

At the point $(x_1, x_2) = (0, 2)$ this becomes

$$\nabla(s) = (2\mathrm{d} x_1 - 2\mathrm{d} x_2 + (\mathrm{d} x_1 + \mathrm{d} x_2)) s_1$$
$$+ (2(\mathrm{d} x_1 + \mathrm{d} x_2) - 2\mathrm{d} x_1 + (\mathrm{d} x_1 - \mathrm{d} x_2)) s_2$$
$$= (3\mathrm{d} x_1 - \mathrm{d} x_2) s_1 + (\mathrm{d} x_1 + \mathrm{d} x_2) s_2.$$

For the vector field

$$X = \frac{\partial}{\partial x_1},$$

we get that

$$\nabla_{\left(\frac{\partial}{\partial x_1}\right)}(s) = 3s_1 + s_2.$$

Finally, if we think of a connection ∇ as giving a way to differentiate a section s, then the covariant derivative with respect to a curve σ is a way for differentiating the section along the curve. This intuition is important for the next section.

18.4. Parallel Transport: Why Connections Are Called Connections

Let us return to the original problem of trying to understand the meaning of

$$s(q) - s(p)$$

(since a derivative should be some type of $\lim_{q \to p}(s(q)-s(p))/(q-p)$), where $s : M \to E$ is a section from the base manifold M to the vector bundle E. The difficulty is that $s(q)$ is a vector in the vector space E_q, while $s(p)$ is a vector in the completely different vector space E_p. We want somehow to use our connection ∇ to compare vectors in E_q with vectors in E_p. This we will do, provided we fix a curve σ in the base manifold M going from the point p to the point q. Let

$$v_q \in E_q.$$

We want to be able to "move" this vector v_q to a vector $v_p \in E_p$. The key lies in the concept of parallel transport along the curve σ.

Definition 18.4.1. *A section s is* parallel *with respect to a curve σ if at all points on the curve, we have*

$$\nabla_{\sigma'}s = 0.$$

(Here σ' denotes the tangent vectors of the curve σ.)

Let us quickly discuss why we are using the word "parallel." In one-variable calculus, a function $f(x)$ is a constant function if and only if its derivative is zero. In the world of vector bundles, after fixing a connection ∇, the analog of a function is a section, and the analog of the derivative of a function is $\nabla_{\sigma'}$. Thus a section s being parallel with respect to a curve σ is the vector bundle analog of a constant function.

Definition 18.4.2. *The* parallel transport *of a vector $v_q \in E_q$ along a curve σ from the point p to q is the value*

$$s(p) \in E_p,$$

where s is a section parallel with respect to σ such that

$$s(q) = v_q.$$

This is just applying the idea that a parallel section s is the vector bundle analog of constant functions. This is why the vector $s(p)$ in E_p should be the vector corresponding to the original vector v_q in E_q.

It is not obvious that such parallel transports exist, though. The following theorem states that they do. (The key for the proof is that we reduce the existence of such parallel transports to solving a system of ordinary differential equations, which can always be done.)

Theorem 18.4.1. *Let ∇ be a connection for a rank k vector bundle E on a manifold M. Let σ be a curve on M going through a point $q \in M$. Then, given any vector $v \in E_q$, there is a unique section s that is parallel with respect to σ such that $s(q) = v$.*

Proof. Let $s_1, \ldots, s_k$ be a basis for our vector bundle. We are given the vector $v \in E_q$. This means that there are numbers $v_1, \ldots, v_k$ such that

$$v = v_1 s_1(q) + \cdots + v_k s_k(q).$$

Let $x_1, \ldots, x_n$ be our local coordinates for M. Label $q \in M$ as

$$q = (q_1, \ldots, q_n).$$

Let our curve σ be described via

$$\sigma(t) = (x_1(t), \ldots, x_n(t)).$$

We can assume that

$$\sigma(0) = (x_1(0), \ldots, x_n(0)) = (q_1, \ldots, q_n) = q.$$

We know that the tangent vector to σ is

$$\sigma'(t) = \left(\frac{dx_1(t)}{dt} \right) \frac{\partial}{\partial x_1} + \cdots + \left(\frac{dx_n(t)}{dt} \right) \frac{\partial}{\partial x_n}.$$

We know for any section

$$s = f_1 s_1 + \cdots + f_k s_k$$

that

$$\nabla_{\sigma'}(s) = \left(\sum_{i=1}^{k} f_i \omega_{i1}(\sigma') + df_1(\sigma') \right) s_1 + \cdots + \left(\sum_{i=1}^{k} f_i \omega_{ik}(\sigma') + df_k(\sigma') \right) s_k.$$

We must find functions $f_1, \ldots, f_k$ such that

$$f_1(q) = v_1, \ldots, f_k(q) = v_q$$

and such that, for the tangent vectors σ' along our curve σ, we have

$$\nabla_{\sigma'}(s) = 0.$$

But this means we must find the functions f_i such that

$$\sum_{i=1}^{k} f_i \omega_{i1}(\sigma') + \mathrm{d} f_1(\sigma') = 0$$

$$\vdots$$

$$\sum_{i=1}^{k} f_i \omega_{ik}(\sigma') + \mathrm{d} f_k(\sigma') = 0$$

Now

$$\mathrm{d} f_i(\sigma') = \left(\frac{\partial f_i}{\partial x_1} \cdot \mathrm{d} x_1 + \cdots + \frac{\partial f_i}{\partial x_n} \cdot \mathrm{d} x_n \right)(\sigma')$$

$$= \left(\frac{\partial f_i}{\partial x_1} \cdot \mathrm{d} x_1 + \cdots + \frac{\partial f_i}{\partial x_n} \cdot \mathrm{d} x_n \right) \left(\sum_{j=1}^{n} \left(\frac{\mathrm{d} x_j(t)}{\mathrm{d} t} \right) \frac{\partial}{\partial x_j} \right)$$

$$= \frac{\partial f_i}{\partial x_1} \cdot \frac{\mathrm{d} x_1(t)}{\mathrm{d} t} + \cdots + \frac{\partial f_i}{\partial x_n} \cdot \frac{\mathrm{d} x_n(t)}{\mathrm{d} t}$$

$$= \frac{\mathrm{d}}{\mathrm{d} t} (f_i(x_1(t), \ldots, x_n(t))).$$

The key for us is that $\frac{\mathrm{d}}{\mathrm{d} t} (f_i(x_1(t), \ldots, x_n(t)))$ is an ordinary derivative.

Thus to find the f_i so that $\nabla_{\sigma'}(s) = 0$ comes down to solving by following the system of k ordinary differential equations. We need

$$\frac{\mathrm{d}}{\mathrm{d} t} (f_j(x_1(t), \ldots, x_n(t)))$$

to equal

$$-\sum_{i=1}^{k} f_i \omega_{ij} \left(\left(\frac{\mathrm{d} x_1(t)}{\mathrm{d} t} \right) \frac{\partial}{\partial x_1} + \cdots + \left(\frac{\mathrm{d} x_n(t)}{\mathrm{d} t} \right) \frac{\partial}{\partial x_n} \right),$$

for all $j = 1, \ldots k$, with initial conditions $f_1(q) = v_1, \ldots, f_k(q) = v_q$. The left-hand side of the preceding consists of ordinary derivatives while the right-hand side is made up of our unknown functions f_i and known functions. Such systems always have unique solutions, as seen in almost all books on ordinary differential equations.   □

   The uniqueness of parallel transports is important. Start with our fixed curve σ in the base manifold M, starting at the point p and ending at q (Figure 18.4).

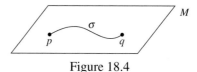

Figure 18.4

The vector bundle E over the curve σ is (Figure 18.5)

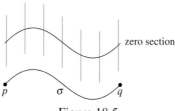

Figure 18.5

(Here we are drawing E as a dimension one vector bundle, which is of course not necessarily the case.) Fix a vector $v_q \in E_q$. There are (infinitely) many sections s over σ such that $s(q) = v_q$.

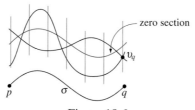

Figure 18.6

But there is only one section s that is parallel with respect to σ with $s(q) = v_q$. Thus the parallel transport of v_q to the point $s(p)$ is unique.

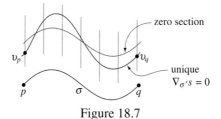

Figure 18.7

Finally, why are connections called connections? Parallel transports exist, letting us move a vector $v \in E_q$ to vectors in other fibers. Thus having a connection ∇ allows us to "connect," or link, vectors in any E_q with vectors in any other E_p.

18.5. Appendix: Tensor Products of Vector Spaces

18.5.1. A Concrete Description

Let V be an n-dimensional vector space and W be an m-dimensional vector space. We want to construct a new vector space, denoted by $V \otimes W$, whose dimension will be nm. We call this new vector space $V \otimes W$ the *tensor product* of V and W.

Let $v_1, \ldots, v_n$ be a basis for V and let $w_1, \ldots, w_m$ be a basis for W. Right now, take as our basis for $V \otimes W$ to be all possible

$$v_i \otimes w_j.$$

Then define the elements of $V \otimes W$ to be all possible

$$\sum_{j=1}^{m} \sum_{i=1}^{n} a_{ij} v_i \otimes w_j,$$

with the a_{ij} being numbers. For example, if $n = 2$ and $m = 3$, then an element of $V \otimes W$ will be of the form

$$a_{11} v_1 \otimes w_1 + a_{12} v_1 \otimes w_2 + a_{13} v_1 \otimes w_3 + a_{21} v_2 \otimes w_1 + a_{22} v_2 \otimes w_2 + a_{23} v_2 \otimes w_3.$$

We now add a few extra requirements onto the algebraic structure of $V \otimes W$. We require, for all vectors $v, \tilde{v} \in V$, $w, \tilde{w} \in W$ and scalars λ, that

$$(v + \tilde{v}) \otimes w = v \otimes w + \tilde{v} \otimes w,$$

$$v \otimes (w + \tilde{w}) = v \otimes w + v \otimes \tilde{w}$$

and

$$\lambda(v \otimes w) = (\lambda v) \otimes w = v \otimes (\lambda w).$$

These rules allow us to interpret any $v \otimes w$ as a linear combination of various $v_i \otimes w_j$, as described in

Theorem 18.5.1. *Suppose*

$$v = \alpha_1 v_1 + \cdots + \alpha_n v_n \in V$$

and

$$w = \beta_1 w_1 + \cdots + \beta_m w_m \in W.$$

Then

$$v \otimes w = \sum_{i=1,j=1}^{i=n,j=m} \alpha_i \beta_j v_i \otimes w_j.$$

(The proof is one of the exercises at the end of the chapter.)
For example, we have

$$(2v_1 + 3v_2) \otimes (4w_1 + w_2) = (2v_1 + 3v_2) \otimes (4w_1) + (2v_1 + 3v_2) \otimes w_2$$

$$= (2v_1) \otimes (4w_1) + (2v_1) \otimes w_2$$

$$+ (3v_2) \otimes (4w_1) + (3v_2) \otimes w_2$$

$$= 8v_1 \otimes w_1 + 2v_1 \otimes w_2$$

$$+ 12v_2 \otimes w_1 + 3v_2 \otimes w_2.$$

18.5.2. Alternating Forms as Tensors

Let V be a vector space. Earlier we defined alternating k-forms $\Lambda^k V$. These can also be defined in term of tensors. We start with our reinterpretation of 2-forms, $\Lambda^2 V$, which we define to be the subspace of $V \otimes V$ generated by all elements of the form

$$v \otimes w - w \otimes v,$$

for any $v, w \in V$. We set

$$v \wedge w = v \otimes w - w \otimes v.$$

Note that

$$v \wedge w = v \otimes w - w \otimes v = -(w \otimes v - v \otimes w) = -w \wedge v.$$

Thus

$$v \wedge v = -v \wedge v = 0.$$

Given a basis for V, we have a natural basis for $\Lambda^2 V$:

Theorem 18.5.2. *Let $v_1, \ldots, v_n$ be a basis for V. Then a basis for $\Lambda^2 V$ will be all*

$$v_i \wedge v_j,$$

for $1 \leq i < j \leq n$.

The proof is left for the exercises.

Let us see how this works. Suppose V is a two-dimensional vector space, with basis v_1 and v_2. Let us write

$$(2v_1 + 3v_2) \wedge (4v_1 + v_2)$$

in terms of
$$v_1 \wedge v_2.$$

By definition
$$(2v_1 + 3v_2) \wedge (4v_1 + v_2) = (2v_1 + 3v_2) \otimes (4v_1 + v_2) - (4v_1 + v_2) \otimes (2v_1 + 3v_2).$$

We know from the preceding that
$$(2v_1 + 3v_2) \otimes (4v_1 + v_2) = 8v_1 \otimes v_1 + 2v_1 \otimes v_2 + 12v_2 \otimes v_1 + 3v_2 \otimes v_2.$$

Via calculation, we have
$$(4v_1 + v_2) \otimes (2v_1 + 3v_2) = 8v_1 \otimes v_1 + 12v_1 \otimes v_2 + 2v_2 \otimes v_1 + 3v_2 \otimes v_2.$$

Then
$$
\begin{aligned}
(2v_1 + 3v_2) \wedge (4v_1 + v_2) &= 8v_1 \otimes v_1 + 2v_1 \otimes v_2 + 12v_2 \otimes v_1 + 3v_2 \otimes v_2 \\
&\quad - (8v_1 \otimes v_1 + 12v_1 \otimes v_2 + 2v_2 \otimes v_1 + 3v_2 \otimes v_2) \\
&= -10v_1 \otimes v_2 + 10v_2 \otimes v_1 \\
&= (-10)(v_1 \otimes v_2 - v_2 \otimes v_1) \\
&= -10v_1 \wedge v_2.
\end{aligned}
$$

The space $\Lambda^3 V$ will be the subspace of $V \otimes V \otimes V$ spanned by
$$u \otimes v \otimes w - v \otimes u \otimes w - w \otimes v \otimes u - u \otimes w \otimes v + w \otimes u \otimes v + v \otimes w \otimes u.$$

Each of these vectors we denote by
$$u \wedge v \wedge w.$$

Thus $u \wedge v \wedge w$ is the linear combination of the six possible tensor products of $u, v, w \in V$, with coefficients ± 1 depending on whether the permutation is even or odd. We have

Theorem 18.5.3. *Let $v_1, \ldots, v_n$ be a basis for V. Then a basis for $\Lambda^3 V$ will be all*
$$v_i \wedge v_j \wedge v_k,$$
for $1 \le i < j < k \le n$.

The proof is left for the exercises.

In general, the alternating k-forms $\Lambda^k V$ will be the subspace of $V \otimes \cdots \otimes V$ (V tensored with itself k times) generated by the sum of all tensor products of k vectors under all possible permutations, with coefficients ± 1 depending on whether the permutation is even or odd.

Theorem 18.5.4. *Let $v_1, \ldots, v_n$ be a basis for V. Then a basis for $\Lambda^k V$ will be all*

$$v_{i_1} \wedge \cdots \wedge v_{i_k},$$

for $1 \leq i_1 < \cdots < i_k \leq n$.

18.5.3. Homogeneous Polynomials as Symmetric Tensors

Besides alternating tensors, there is another natural subspace of any tensor space $V \otimes \cdots \otimes V$, the symmetric tensors. In this section we will define symmetric tensors and then see how they can be easily interpreted as homogeneous polynomials. This suggests how flexible tensor notation can be, as it can be used to capture not only the language of differential forms but also the language of much of algebra.

Let V be a vector space. Define $S^2(V)$ (whose elements are called the symmetric two-tensors) to be the subspace of $V \otimes V$ generated by all elements of the form

$$v \otimes w + w \otimes v,$$

for any $v, w \in V$. We set

$$v \odot w = v \otimes w + w \otimes v.$$

Theorem 18.5.5. *Let $v_1, \ldots, v_n$ be a basis for V. Then a basis for $S^2(V)$ will be all*

$$v_i \odot v_j,$$

for $1 \leq i \leq j \leq n$.

The proof is left for the exercises.

In analog to alternating k-forms, the symmetric k-forms $S^k(V)$ will be the subspace of $V \otimes \cdots \otimes V$ (V tensored with itself k times) generated by the sum of all tensor products of k vectors under all possible permutations.

Theorem 18.5.6. *Let $v_1, \ldots, v_n$ be a basis for V. Then a basis for $S^k V$ will be all*

$$v_{i_1} \odot \cdots \odot v_{i_k},$$

for $1 \leq i_1 \leq \cdots \leq i_k \leq n$.

The proof is also left for the exercises.

Here we are generalizing the use of $\odot$ to mean that $v_{i_1} \odot \cdots \odot v_{i_k}$ should be interpreted to be the sum of all tensor products of k vectors under all possible permutations.

Now to include homogeneous polynomials. Homogeneous polynomials in the two variables x_1 and x_2 are simply those polynomials that are the sum of monomials of the same degree. Thus

$$x_1 x_2 + x_2^2$$

is homogeneous of degree two, since its two monomials $x_1 x_2$ and x_2^2 each have degree two, while

$$x_1 x_2 + x_2^3$$

is not homogeneous, since x_2^3 has degree three. In two variables, all degree two homogeneous polynomials are of the form

$$ax_1^2 + bx_1 x_2 + cx_2^2,$$

while all degree three homogeneous polynomials are of the form

$$ax_1^3 + bx_1^2 x_2 + cx_1 x_2^2 + dx_2^3.$$

Similar definitions hold for three variables. For example, in three variables, all degree two homogeneous polynomials are of the form

$$ax_1^2 + bx_1 x_2 + cx_1 x_3 + dx_2^2 + ex_2 x_3 + fx_3^2.$$

Now to link with symmetric tensors. Let V be a two-dimensional vector space with basis x_1 and x_2. Since $S^2(V)$ has basis $x_1 \odot x_1$, $x_1 \odot x_2$ and $x_2 \odot x_2$, we know that all symmetric two-tensors are of the form

$$ax_1 \odot x_1 + bx_1 \odot x_2 + cx_2 \odot x_2.$$

Each such symmetric two-tensor can of course be effortlessly thought of as the polynomial $ax_1^2 + bx_1 x_2 + cx_2^2$.

In general, if V has basis $x_1, \ldots, x_n$, we associate each

$$x_{i_1} \odot \cdots \odot x_{i_k}$$

to the polynomial

$$x_{i_1} \cdots x_{i_k}.$$

What is important is that a change of basis on V will correspond to a homogeneous linear change of coordinates for the corresponding polynomials.

18.5.4. *Tensors as Linearizations of Bilinear Maps*

So far in this Appendix we have been emphasizing how to compute and construct tensor spaces. Here we will give a more intrinsic approach. The natural maps between vector spaces are linear transformations. But a number of times in this text we have looked not at linear maps but instead at bilinear maps. Given three vector spaces U, V, and W, recall that a bilinear map is a map

$$B : U \times V \to W$$

such that, for all vectors $u_1, u_2 \in U$, $v_1, v_2 \in V$ and scalars $\alpha_1, \alpha_2 \in \mathbb{R}$, we have

$$B(\alpha_1 u_1 + \alpha_2 u_2, v_1) = \alpha_1 B(u_1, v_1) + \alpha_2 B(u_2, v_1)$$

$$B(u_1, \alpha_1 v_1 + \alpha_2 v_2) = \alpha_1 B(u_1, v_1) + \alpha_2 B(u_1, v_2).$$

At almost the level of fantasy math, it would be great if we could somehow translate this bilinear map into a linear map. Somewhat surprisingly this can be done:

Theorem 18.5.7. *Given any two vector spaces U and V, there exists a third vector space, denoted by $U \otimes V$ and a natural bilinear map*

$$\pi : U \times V \to U \otimes V$$

such that for any other vector space W and any bilinear map $B : U \times V \to W$ there exists a unique linear map

$$b : U \otimes V \to W$$

such that

$$B = b \circ \pi.$$

Thus we always have the following commutative diagram:

$$
\begin{array}{ccc}
U \times V & \overset{B}{\to} & W \\
\pi \downarrow & \nearrow b & \\
U \otimes V & &
\end{array}
$$

Here are the actual details of how to link B with b. Let $u_1, \ldots, u_m$ be a basis for U and let $v_1, \ldots, v_n$ be a basis for V. The map

$$\pi : U \times V \to U \otimes V$$

is defined by setting

$$\pi(u, v) = u \otimes v.$$

If

$$u = \alpha_1 u_1 + \cdots + \alpha_m u_m$$

and

$$v = \beta_1 v_1 + \cdots + \beta_n v_n,$$

then

$$\pi(u,v) = (\alpha_1 u_1 + \cdots + \alpha_m u_m) \otimes (\beta_1 v_1 + \cdots + \beta_n v_n)$$
$$= \sum \alpha_i \beta_j u_i \otimes v_j.$$

Any bilinear map $B : U \times V \to W$ is determined by its values

$$B_{ij} = B(u_i, v_j).$$

Define the map $b : U \otimes V \to W$ by setting

$$B_{ij} = b(u_i \otimes v_j).$$

We want to show that

$$B(u,v) = b(u \otimes v).$$

We have

$$B(u,v) = B(\alpha_1 u_1 + \cdots + \alpha_m u_m, \beta_1 v_1 + \cdots + \beta_n v_n)$$
$$= \sum \alpha_i B(u_i, \beta_1 v_1 + \cdots + \beta_n v_n)$$
$$= \sum \alpha_i \beta_j B(u_i, v_j)$$
$$= \sum \alpha_i \beta_j b(u_i \otimes v_j)$$
$$= b((\alpha_1 u_1 + \cdots + \alpha_m u_m) \otimes (\beta_1 v_1 + \cdots + \beta_n v_n))$$
$$= b(u \otimes v),$$

as desired.

18.6. Exercises

Exercise 18.6.1. *Describe* $\mathbb{R}^3$ *as a rank-two trivial bundle with base manifold* $\mathbb{R}$.

Exercise 18.6.2. *Describe* $\mathbb{R}^3$ *as a rank-one trivial bundle with base manifold* $\mathbb{R}^2$.

Exercise 18.6.3. *Let E be a rank-two bundle with basis of sections s_1 and s_2 on a surface M, which has coordinates x_1 and x_2. Let the connection matrix be*

$$\omega = \begin{pmatrix} dx_1 + 2dx_2 & dx_1 + dx_2 \\ dx_1 - dx_2 & dx_1 \end{pmatrix}.$$

For

$$s = (x_1^2)s_1 + (x_1 x_2)s_2,$$

find

$$\nabla(s).$$

Exercise 18.6.4. *Let E be a rank-two bundle with basis of sections s_1 and s_2 on a surface M, which has coordinates x_1 and x_2. Let the connection matrix be*

$$\omega = \begin{pmatrix} \mathrm{d}x_1 & \mathrm{d}x_1 + \mathrm{d}x_2 \\ \mathrm{d}x_2 & \mathrm{d}x_1 \end{pmatrix}.$$

Let

$$s = (x_2\mathrm{d}x_1) \otimes s_1 + (x_2\mathrm{d}x_2) \otimes s_2 \in \Lambda^1(M) \otimes \Gamma(E).$$

Compute

$$\nabla(s).$$

Exercise 18.6.5. *Let E be the trivial rank-two bundle with base manifold $\mathbb{R}^2$ with trivial connection. Let x and y be the coordinates for $\mathbb{R}^2$. For the section*

$$s(x,y) = \begin{pmatrix} x + y^2 \\ x^3 \end{pmatrix}$$

and curve

$$\sigma(t) = (t, t^2),$$

find

$$\nabla_{\sigma'(0)} s.$$

Exercise 18.6.6. *Let E be the trivial rank-two bundle with base manifold $\mathbb{R}^2$ with trivial connection. Let*

$$q = (2, 4) \in \mathbb{R}^2$$

and

$$v = \begin{pmatrix} 3 \\ 4 \end{pmatrix}.$$

Parallel transport v to a vector in $E_{(0,0)}$ along the curve

$$\sigma(t) = (t, t^2).$$

Exercise 18.6.7. *Do the same as in the previous problem, but now parallel transport the vector v along the path*

$$\sigma(t) = (t, 2t).$$

Exercise 18.6.8. *Show that the zero section of a vector bundle E (i.e., $s(p) = 0$ for any $p \in M$) is parallel to any curve $\sigma(t) \subset M$, with respect to any connection.*

Exercise 18.6.9. *Let V have basis vectors v_1 and v_2 and let W have basis vectors w_1, w_2, and w_3. Write*

$$(2v_1 + 3v_2) \otimes (4w_1 + w_2 + 5w_3)$$

in terms of the basis formed from the various $v_i \otimes w_j$.

Exercise 18.6.10. *Let $v_1, \ldots, v_n$ be a basis for V and let $w_1, \ldots, w_m$ be a basis for W. Suppose*

$$v = \alpha_1 v_1 + \cdots + \alpha_n v_n \in V$$

and

$$w = \beta_1 w_1 + \cdots + \beta_m w_m \in W.$$

Then show that

$$v \otimes w = \sum_{i=1, j=1}^{i=n, j=m} \alpha_i \beta_j v_i \otimes w_j.$$

Exercise 18.6.11. *Let v_1, v_2, v_3 be a basis for V. Write*

$$(3v_1 + 2v_2 + 4v_3) \wedge (v_1 - v_2 + v_3)$$

in terms of the various $v_i \wedge v_j$, with $i < j$.

Exercise 18.6.12. *Let $v_1, \ldots, v_n$ be a basis for V. Show that all*

$$v_i \wedge v_j,$$

for $1 \leq i < j \leq n$, form a basis for $\Lambda^2 V$.

Exercise 18.6.13. *Let $v_1, \ldots, v_n$ be a basis for V. Show that all*

$$v_i \wedge v_j \wedge v_k,$$

for $1 \leq i < j < k \leq n$, form a basis for $\Lambda^3 V$.

Exercise 18.6.14. *Let V have basis vectors v_1 and v_2. Write*

$$(2v_1 + 3v_2) \odot (4v_1 + v_2 + 5v_3)$$

in terms of the basis formed from the various $v_i \odot v_j$.

Exercise 18.6.15. *Let* $v_1, \ldots, v_n$ *be a basis for* V. *Show that all*

$$v_i \odot v_j,$$

for $1 \le i \le j \le n$, *form a basis for* $S^2(V)$.

Exercise 18.6.16. *Let* $v_1, \ldots, v_n$ *be a basis for* V. *Show that all*

$$v_{i_1} \odot \cdots \odot v_{i_k},$$

for $1 \le i_1 \le \cdots \le i_k \le n$, *form a basis for* $S^k V$.

19

Curvature

Summary: The goal of this chapter is to define the curvature of a vector bundle. Curvature will depend on a choice for a connection for the vector bundle. If a connection provides a method for differentiating sections of a vector bundle, then the curvature can be interpreted as taking the second derivative of a section.

19.1. Motivation

The study of curvature lies at the heart of much of geometry, if not most of mathematics. Originally, curvature captured the rate of change of tangent vectors of a manifold. Since tangent vectors in turn are captured by first derivative type information, we should expect curvature to be rates of change of derivatives and hence involve second derivatives.

Probably most people's introduction to these ideas is in the study of concavity properties of single-variable functions in beginning calculus.

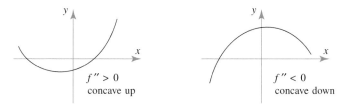

Figure 19.1

Here (Figure 19.1) a curve $y = f(x)$ is concave up if $f''(x) > 0$ and concave down if $f''(x) < 0$.

The second standard place to see curvature is in a multivariable calculus course. Starting with a plane curve $(x(t), y(t))$ (Figure 19.2),

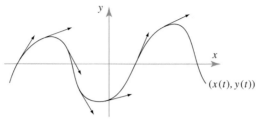

Figure 19.2

the curvature is the rate of change of the unit tangent vector with respect to arclength. The actual curvature is shown to be

$$\frac{|x'y'' - y'x''|}{\left((x')^2 + (y')^2\right)^{3/2}}.$$

Note that the formulas already are becoming a bit complicated. This is a recurring theme, namely, that formulas exist but are difficult to understand truly.

Thinking of curvature as second derivative type information suggests the following approach for defining curvature for vector bundles. Start by fixing a connection ∇, which as we saw in the last chapter is a method for defining derivatives of sections. Then curvature should be something like

$$\text{Curvature} = \nabla \circ \nabla,$$

or, in other words, the connection of the connection. Now to make sense of this intuition.

19.2. Curvature and the Curvature Matrix

Let E be a rank k vector bundle with base manifold M. Fix a connection ∇ on E. Motivated by the belief that curvature should be a second derivative, we have

Definition 19.2.1. *The curvature for the connection ∇ is the map*

$$\nabla^2 : \Gamma(E) \to \Lambda^2(M) \otimes \Gamma(E)$$

defined by setting

$$\nabla^2(s) = \nabla(\nabla(s)).$$

We can capture the curvature by a $k \times k$ matrix of 2-forms. Fix a local frame of sections $s_1, \ldots, s_k$ for E. Then we have the associated $k \times k$ connection matrix of 1-forms $\omega = (\omega_{ij})$.

Definition 19.2.2. *The* curvature matrix *with respect to a connection* ∇ *and local frame* $s_1, \ldots, s_k$ *is*

$$\Omega = \omega \wedge \omega - d\omega.$$

Before showing how to use the connection matrix Ω to compute curvature, let us look at an example. Suppose that E has rank two, and that the connection matrix is

$$\omega = \begin{pmatrix} \omega_{11} & \omega_{12} \\ \omega_{21} & \omega_{22} \end{pmatrix}.$$

Then

$$\Omega = \omega \wedge \omega - d\omega$$

$$= \begin{pmatrix} \omega_{11} & \omega_{12} \\ \omega_{21} & \omega_{22} \end{pmatrix} \wedge \begin{pmatrix} \omega_{11} & \omega_{12} \\ \omega_{21} & \omega_{22} \end{pmatrix} - \begin{pmatrix} d\omega_{11} & d\omega_{12} \\ d\omega_{21} & d\omega_{22} \end{pmatrix}$$

$$= \begin{pmatrix} \omega_{11} \wedge \omega_{11} + \omega_{12} \wedge \omega_{21} & \omega_{11} \wedge \omega_{12} + \omega_{12} \wedge \omega_{22} \\ \omega_{21} \wedge \omega_{11} + \omega_{22} \wedge \omega_{21} & \omega_{21} \wedge \omega_{12} + \omega_{22} \wedge \omega_{22} \end{pmatrix} - \begin{pmatrix} d\omega_{11} & d\omega_{12} \\ d\omega_{21} & d\omega_{22} \end{pmatrix}$$

$$= \begin{pmatrix} \omega_{11} \wedge \omega_{11} + \omega_{12} \wedge \omega_{21} - d\omega_{11} & \omega_{11} \wedge \omega_{12} + \omega_{12} \wedge \omega_{22} - d\omega_{12} \\ \omega_{21} \wedge \omega_{11} + \omega_{22} \wedge \omega_{21} - d\omega_{21} & \omega_{21} \wedge \omega_{12} + \omega_{22} \wedge \omega_{22} - d\omega_{22} \end{pmatrix}.$$

The link between this matrix of 2-forms and curvature lies in the following. A section S of E is any

$$S = f_1 s_1 + \cdots + f_k s_k,$$

for real-valued functions $f_1, \ldots, f_k$. As in the last chapter, let

$$f = (\,f_1 \quad \cdots \quad f_k\,)$$

and let s denote the corresponding column vector of sections with entries $s_1, \ldots, s_k$. Then

$$S = f \cdot s.$$

The key is the following theorem, which we will prove in the next section:

Theorem 19.2.1.

$$\nabla^2(S) = f \cdot \Omega \cdot s.$$

Thus for a rank-two bundle, we have

$$\nabla^2(S) = f \cdot \Omega \cdot s$$

$$= (\,f_1 \quad f_2\,) \begin{pmatrix} \Omega_{11} & \Omega_{12} \\ \Omega_{21} & \Omega_{22} \end{pmatrix} \begin{pmatrix} s_1 \\ s_2 \end{pmatrix}$$

$$= (f_1 \Omega_{11} + f_2 \Omega_{21}) s_1 + (f_1 \Omega_{12} + f_2 \Omega_{22}) s_2.$$

19.3. Deriving the Curvature Matrix

Our goal is to prove that

$$\nabla^2(S) = f \cdot \Omega \cdot s.$$

We will be using that it is always the case that

$$d^2 f = 0.$$

Also, for a section s and an m-form τ, we have

$$\nabla(\tau \otimes s) = (-1)^m d\tau \otimes s + \tau \wedge \nabla(s).$$

We have

$$
\begin{aligned}
\nabla^2(S) &= \nabla(\nabla(S)) \\
&= \nabla(\nabla(f \cdot s)) \\
&= \nabla(df \cdot s + f \cdot \omega \cdot s) \\
&= \nabla(df \cdot s) + \nabla(f \cdot \omega \cdot s).
\end{aligned}
$$

Now, df is a row matrix of 1-forms. Thus

$$
\begin{aligned}
\nabla(df \cdot s) &= -d(df) \cdot s + df \wedge \nabla(s) \\
&= df \wedge \omega \cdot s,
\end{aligned}
$$

since $d(df) = d^2 f = 0$. We can think of $f \cdot \omega$ as a row matrix of 1-forms. Using that

$$d(f \cdot \omega) = df \wedge \omega + f \cdot d\omega,$$

we have

$$
\begin{aligned}
\nabla(f \cdot \omega \cdot s) &= -d(f \cdot \omega) \cdot s + f \cdot \omega \wedge \nabla(s) \\
&= -df \wedge \omega \cdot s - f \cdot d\omega \cdot s + f \cdot \omega \wedge \omega \cdot s.
\end{aligned}
$$

Putting all of this together, we have

$$\nabla^2(S) = f \cdot (\omega \wedge \omega - d\omega) \cdot s = f \cdot \Omega \cdot s,$$

as desired.

Some of the exercises ask for you to give more coordinate-dependent proofs.

19.4. Exercises

Exercise 19.4.1. *Let E be a rank-two vector bundle with base manifold $\mathbb{R}^2$, with coordinates x_1, x_2. Suppose we have the connection matrix*

$$\begin{pmatrix} x_2^2 \mathrm{d}x_1 + \mathrm{d}x_2 & x_1 \mathrm{d}x_1 + \mathrm{d}x_2 \\ x_2 \mathrm{d}x_1 - \mathrm{d}x_2 & \mathrm{d}x_1 + x_1 \mathrm{d}x_2 \end{pmatrix},$$

with respect to a basis of sections s_1 and s_2 for E. Compute the curvature matrix Ω.

Exercise 19.4.2. *Using the same notation as in the previous problem, consider a section*

$$S = (x_1 x_2) s_1 + (x_1 + x_2) s_2.$$

Find the curvature of this section.

Exercise 19.4.3. *Let E be a rank-two vector bundle with base manifold $\mathbb{R}^2$, with coordinates x_1, x_2. Suppose we have the connection matrix*

$$\begin{pmatrix} x_2 \mathrm{d}x_1 + (x_1 + x_2) \mathrm{d}x_2 & x_1 \mathrm{d}x_1 + \mathrm{d}x_2 \\ x_2 \mathrm{d}x_1 + \mathrm{d}x_2 & \mathrm{d}x_1 - x_1 \mathrm{d}x_2 \end{pmatrix},$$

with respect to a basis of sections s_1 and s_2 for E. Consider the section

$$S = (x_1 - x_2) s_1 + (x_1^2 + x_2) s_2.$$

Find the curvature of this section.

Exercise 19.4.4. *Let E be a rank k bundle over a one-dimensional base manifold M. Let ω be a connection matrix for E with respect to some basis of sections for E. Compute the curvature of any section. (Hint: Since so little information is given, the actual answer cannot be that hard.)*

Exercise 19.4.5. *Let E be a trivial bundle on a manifold M, with trivial connection. Find the corresponding curvature matrix.*

Exercise 19.4.6. *Let*

$$f = \begin{pmatrix} f_1 & f_2 \end{pmatrix}$$

be a row vector of functions and

$$\omega = \begin{pmatrix} \omega_{11} & \omega_{12} \\ \omega_{21} & \omega_{22} \end{pmatrix}$$

a matrix of 1-forms. Prove that

$$\mathrm{d}(f \cdot \omega) = \mathrm{d}f \wedge \omega + f \cdot \mathrm{d}\omega.$$

The next two exercises are deriving the formula for the curvature matrix via local calculations.

Exercise 19.4.7. *Let E be a rank-two vector bundle over a base manifold M. Let s_1 and s_2 be a basis of sections. Suppose that there is a connection matrix*

$$\omega = \begin{pmatrix} \omega_{11} & \omega_{12} \\ \omega_{21} & \omega_{22} \end{pmatrix}.$$

For a section $S = f_1 s_1 + f_2 s_2$, we know that

$$\nabla(S) = (df_1 + f_1\omega_{11} + f_2\omega_{21})s_1 + (df_2 + f_1\omega_{12} + f_2\omega_{22})s_2.$$

Explicitly calculate the curvature $\nabla^2(S)$ and then show that it agrees with our matrix formulation of $f \cdot \Omega \cdot s$.

Exercise 19.4.8. *Let E be a rank-k vector bundle over a base manifold M. Let $s_1, \ldots, s_k$ be a basis of sections. Suppose that there is a connection matrix*

$$\omega = \begin{pmatrix} \omega_{11} & \cdots & \omega_{1k} \\ & \vdots & \\ \omega_{k1} & \cdots & \omega_{kk} \end{pmatrix}.$$

For a section $S = f_1 s_1 + \cdots + f_k s_k$, we know that

$$\nabla(S) = \sum_{i=1}^{k} \left(df_i + \sum_{j=1}^{k} f_j \omega_{ji} \right) s_i.$$

Explicitly calculate the curvature $\nabla^2(S)$ and then show that it agrees with our matrix formulation of $f \cdot \Omega \cdot s$.

20

Maxwell via Connections and Curvature

Summary: We finally pull together all the themes of this book, formulating Maxwell's equations in terms of connections and curvature. It is this formulation that will allow deep generalizations in the next chapter.

20.1. Maxwell in Some of Its Guises

At the beginning of this book we wrote down Maxwell's equations:

$$\text{div}(E) = \rho$$

$$\text{curl}(E) = -\frac{\partial B}{\partial t}$$

$$\text{div}(B) = 0$$

$$c^2 \, \text{curl}(B) = j + \frac{\partial E}{\partial t}.$$

Here

$$E = (E_1(x,y,z,t), E_2(x,y,z,t)), E_3(x,y,z,t))$$

is the electric field,

$$B = (B_1(x,y,z,t), B_2(x,y,z,t)), B_3(x,y,z,t))$$

is the magnetic field, the function ρ is the charge density, and

$$j = (J_1(x,y,z,t), J_2(x,y,z,t), J_3(x,y,z,t))$$

is the current. We showed that there always exist a scalar potential function $\phi(x,y,z,t)$ and a vector potential $A = (A_1(x,y,z,t), A_2(x,y,z,t), A_3(x,y,z,t))$

such that

$$E = -\nabla\phi - \frac{\partial A}{\partial t}$$

$$B = \nabla \times A.$$

We then recast these equations into the language of differential forms. We defined the electromagnetic 2-form to be

$$F = E_1 dx \wedge dt + E_2 dy \wedge dt + E_3 dz \wedge dt$$
$$+ B_1 dy \wedge dz + B_2 dz \wedge dx + B_3 dx \wedge dy$$

and the potential 1-form to be

$$A = -\phi dt + A_1 dx + A_2 dy + A_3 dz,$$

with the relation

$$F = dA.$$

Encoding the charge density ρ and the current j as the 1-form

$$J = -\rho dt + J_1 dx + J_2 dy + J_3 dz,$$

we saw that Maxwell's equations could be written as

$$dF = 0$$

$$\star d \star F = J.$$

We further saw that the function

$$L = (\star J) \wedge A + \frac{1}{2}(\star F) \wedge F$$

serves as a Lagrangian for Maxwell's equations, meaning that the Euler-Lagrange equations for finding the critical points of

$$\int_{\mathbb{R}^4} L \, dx\,dy\,dz\,dt$$

are Maxwell's equations.

20.2. Maxwell for Connections and Curvature

We can now recast Maxwell's equations into the language of connections and curvature. The end result will be that the potential one-form will be a connection matrix and the electromagnetic two-form will be the curvature of this connection. This may seem to be a mere reshuffling of definitions. Its importance is that this will create a natural language for deep generalizations that profoundly impact physics and mathematics.

Let E be a trivial real line bundle over $\mathbb{R}^4$. Choose a section s that is nowhere zero. To define a connection, let ω be a 1-form and define the connection for E by setting, for any function f on $\mathbb{R}^4$,

$$\nabla(fs) = f \cdot \omega \cdot s + \mathrm{d}f \cdot s$$

The connection matrix is simply A. Then the curvature of this connection will be

$$F = \omega \wedge \omega - \mathrm{d}\omega$$
$$= -\mathrm{d}\omega,$$

since the wedge of a 1-form with itself is always zero. For convenience, set $A = -\omega$. Then $F = \mathrm{d}A$.

We want to identify the curvature F with the electromagnetic two-form. The curvature F will correspond to the electromagnetic two-form if

$$\mathrm{d}F = 0$$
$$\star \mathrm{d} \star F = J,$$

where J is the one-form encoding charge density. We always have $\mathrm{d}F = 0$ since $F = \mathrm{d}A$ and $d(d)$ is always zero. The extra condition can be phrased either as $\star \mathrm{d} \star F = J$ or, as discussed in Chapter 11, as finding the critical points for the Lagrangian,

$$L = (\star J) \wedge A + \frac{1}{2}(\star F) \wedge F.$$

Thus we can indeed describe Maxwell's equations as follows: Given a charge density and a current, we have a fixed current one-form J. Among all possible connections for the trivial line bundle E over the manifold $M = \mathbb{R}^4$

$$E$$
$$\downarrow$$
$$M,$$

we choose those connections A whose corresponding curvature $F = \mathrm{d}A$ satisfies $\star \mathrm{d} \star F = J$, or those that are critical points for the Lagrangian $L = (\star J) \wedge A + \frac{1}{2}(\star F) \wedge F$.

The key for generalizing is that the field of the force, here the electric and magnetic fields, is described as the curvature of a connection, which we write for emphasis as

$$\boxed{\text{Force} = \text{Curvature}}$$

Further, among all possible connections, the physically significant ones are those whose corresponding curvatures are critical points for a specified Lagrangian. As mentioned earlier, physicists would use the term "gauge" for connections. The corresponding Euler-Lagrange equations for the connections are called Yang-Mills equations.

20.3. Exercises

Exercise 20.3.1. *For the trivial line bundle E over the manifold $M = \mathbb{R}^4$, consider the connection 1-form*

$$\omega = -xz\mathrm{d}t + (x^2 - yt)\mathrm{d}x + (x + zt^2)\mathrm{d}y + (z^2t - y)\mathrm{d}z.$$

Compute the curvature 2-form $F = -\mathrm{d}\omega$ and show

$$\mathrm{d}F = 0$$

$$\star\,\mathrm{d}\star F = -2z\mathrm{d}t - 2z\mathrm{d}y.$$

Compare this to Exercise 2.3.5

Exercise 20.3.2. *For the trivial line bundle E over the manifold $M = \mathbb{R}^4$, consider the connection 1-form*

$$\omega = -\mathrm{d}t + (x - yt)\mathrm{d}x + (x + zt^2)\mathrm{d}y + (z^2t - y)\mathrm{d}z.$$

Compute the curvature 2-form $F = \mathrm{d}A = -\mathrm{d}\omega$. By computing $\mathrm{d}F$ and $\star\,\mathrm{d}\star F$, determine whether the connection A is from an electromagnetic field with

$$E = (tx, xy, z^2)$$
$$B = (y, xy, zt^2)$$
$$\rho = t^2xyz$$
$$j = (x + y, y + z + t, t + x^2z).$$

21

The Lagrangian Machine, Yang-Mills, and Other Forces

Summary: In this final chapter, we generalize our description of Maxwell's equations via connections and curvature to motivate a description of the weak force and the strong force. This, in turn, will motivate Yang-Mills equations, which have had a profound effect not only on physics but also on geometry. In all of these there is the theme that "Force = Curvature."

21.1. The Lagrangian Machine

Here we set up a general framework for understanding forces. (This section is heavily indebted to section 5.13 of Sternberg's *Group Theory and Physics* [63].)

Start with a vector bundle E over a manifold M:

$$E$$
$$\downarrow$$
$$M$$

The current belief is that all forces can be cast into the language of the *Lagrangian Machine*. Specifically, the Lagrangian Machine is a function

$$\mathcal{L} : \text{Connections}(E) \times \Gamma(M, E) \to C^\infty(M),$$

where $\text{Connections}(E)$ is the space of connections on the bundle E, $\Gamma(M, E)$ is the space of sections from M to E, and $C^\infty(M)$ are the smooth functions on M.

Here is how the real world enters this picture. The base manifold M is our world, the world we see and hence should be four-dimensional space-time $\mathbb{R}^4$. The vector bundle E encodes other information. The forces enter the picture by choosing a connection. In electricity and magnetism, we have seen that the connection is a potential for the corresponding force; the actual force will be the curvature of this connection. This will turn out also to hold for the weak force and the strong force. The path of a particle will be a section. Suppose we want to know the path of a particle going from a point $p \in M$ to a point $q \in M$. We know the various forces that exist, allowing us to choose a connection. We know the extra information about the particle at both p and q, which means that we are given points in the fiber of E over both p and q. Then using the function $\mathcal{L}$, choose the section, whose projection to M is the actual path of the particle, by requiring that the function $\mathcal{L}$ has a critical point for this section.

This is precisely the language used in the last chapter to describe Maxwell's equations. Further, our earlier work on describing classical mechanics via Lagrangians can easily be recast into the Lagrangian Machine. (To a large extent, this is precisely what we have done.)

21.2. U(1) Bundles

We first look at a more quantum mechanical approach to the Lagrangian Machine as applied to Maxwell.

We stick with our base manifold being $M = \mathbb{R}^4$. Quantum mechanics, as we saw, is fundamentally a theory over the complex numbers. This motivates replacing our original trivial real line bundle E with a trivial complex line bundle.

Thus we should consider

$$\mathbb{R}^4 \times \mathbb{C}$$
$$\downarrow$$
$$\mathbb{R}^4.$$

Finally, in quantum mechanics the state of a particle is defined only up to a phase factor $e^{i\alpha}$. This suggests that we want to think of our trivial vector bundle as a $U(1)$ bundle, where $U(1)$ is the unitary group of points on the unit circle:

$$U(1) = \{e^{i\alpha} : \alpha \in \mathbb{R}\},$$

with the group action being multiplication.

Hence in an almost line-for-line copying from the last section, we have a trivial complex line bundle over $\mathbb{R}^4$. Choose a section s that is nowhere zero. Let A be a 1-form. Define a connection for E by setting, for any function f on $\mathbb{R}^4$,

$$\nabla(fs) = s \wedge if A + s \wedge \mathrm{d}f.$$

The connection matrix is now iA. (It is traditional for physicists to put in this extra factor of i, possibly so that it is clear we are now dealing with complex line bundles.) Then the curvature of this connection will be

$$F = iA \wedge iA + i\mathrm{d}(A)$$
$$= i\mathrm{d}A,$$

since the wedge of a 1-form with itself is always zero.

As before, the curvature F corresponds to the electromagnetic 2-form if

$$\mathrm{d}F = 0$$
$$\star \mathrm{d} \star F = J,$$

where J is the 1-form encoding charge density and current.

21.3. Other Forces

The modern approach to understanding a force is to attempt to find the appropriate Lagrangian Machine. This involves identifying the correct vector bundle and even more so the correct Lagrangian. Attempts to unify forces become mathematically the attempt to put two seemingly different forces into the same Lagrangian Machine.

There are three known forces besides electricity and magnetism: the weak force, the strong force, and gravity. The weak force, for example, is what causes a lone neutron to split spontaneously into a proton, an electron, and an electron antineutrino. The strong force is what keeps the nucleus of an atom stable, despite being packed with a number of protons, all with positive charge and hence with the desire to be blown apart. And, of course, gravity is what keeps us grounded on Earth.

All these forces have Lagrangian interpretations. In the 1960s and 1970s the weak force was unified with the electromagnetic force. The very language of this unification was in terms of gauge theory, which is the physicist version of connections on a vector bundle. Soon afterward, a common framework, called the standard model, was given to include also the strong force.

Linking these three forces with gravity is still a mystery. The best current thinking falls under the name of string theory. Unfortunately, any type of experimental evidence for string theory seems a long way off. It has, however, generated some of the most beautiful and important mathematics since the 1980s.

Both the weak force and the strong force are viewed as non-Abelian theories. (Abelian is another term for commutative.) Unlike the Lagrangian machine for electromagnetism, for each of these forces the corresponding vector bundles are not trivial complex line bundles but instead bundles of higher rank. For the weak force, the bundle E will be of rank two while for the strong force it will be of rank three. The corresponding transition functions then cannot be made up of invertible one-by-one matrices (complex valued functions) but instead must be matrices. For the weak force, the transition matrices will be in the special unitary group of two-by-two matrices, $SU(2)$, while for the strong force, the transition matrices will be in the special unitary group of three-by-three matrices, $SU(3)$. The standard model has as its transition matrices elements of $U(1) \times SU(2) \times SU(3)$.

Both the weak force and the strong force are quantum mechanical; that, in the language of the Lagrangian Machine, is why the transition functions are in special unitary groups. Gravity also has a Lagrangian interpretation, but the corresponding transition functions seem to have nothing to do with any type of unitary group; that is one way of saying that the theory of gravity is not yet compatible with quantum mechanics.

But for all of these, it is the case that

$$\boxed{\text{Force} = \text{Curvature}}$$

21.4. A Dictionary

It was a long, slow process for physicists to go from the seemingly nonphysical but mathematical meaningfulness of the vector and scalar potentials to the whole machinery of connections and curvature. In fact, at least for the three fields of electromagnetism, the weak force, and the strong force, the rhetoric was in terms of gauges, not connections.

Mathematicians developed the language of connections for vector bundles from questions about curvature. Physicists, on the other hand, developed the language of gauges from questions about forces. It was only in the 1960s and the 1970s that people began to realize that gauges and connections were actually the same. In 1975, Wu and Yang [69] made the dictionary explicit, in the following table, between the physicists' language of gauges and the mathematicians' language of connections, which we reproduce here (since this is lifted from their paper, do not worry very much about what the various symbols mean; we are copying this only to give a flavor as to possible dictionaries):

Gauge field terminology	Bundle terminology
gauge (or global gauge)	principal coordinate bundle
gauge type	principal fiber bundle
gauge potential b_μ^k	connection on a principal fiber bundle
S_{b_a} (see Sec. V)	transition functions
phase factor Φ_{QP}	parallel displacement
field strength $f_{\mu\nu}^k$	curvature
source J_μ^K	?
electromagnetism	connection on a $U(1)$ bundle
isotopic spin gauge field	connection on a SU_2 bundle
Dirac's monopole quantization	classification of $U(1)$ bundle according to first Chern class
electromagnetism without monopole	connection on a trivial $U(1)$ bundle
electromagnetism with monopole	connection on a nontrivial $U(1)$ bundle

Now to break briefly from the standard impersonal writing of math and science. In the early 1980s, as a young graduate student in mathematics, I was concerned with curvature conditions on algebraic vector bundles. Hence I was intensely interested in connections, soon seeing how the abstractions that we have developed here fit naturally into curvature questions. At the same time, it was almost impossible not to hear in the halls of math departments talk about gauges and the links among gauges, bundles, and elementary particles. In fact, I certainly knew that connections were gauges, but for the life of me could not imagine how anyone interested in the real world of physics could have come up with the idea of connections. In the mid 1980s I heard Yang give a talk at Brown University. This is the Yang of the preceding dictionary and, more importantly, one of the prime architects of modern physics, someone who makes up half of the phrase "Yang-Mills," the topic of the next section. In this talk, he spoke about his growing recognition that mathematicians' connections and physicists' gauges were the same. He commented on the fact that it took years for people to realize that these notions were the same, despite the fact that often the mathematicians developing connections and the physicists developing gauge theory worked in the same buildings.[1] Then he said that, for him, it was clear how and why physicists developed the theory of gauges. The surprise for him was that mathematicians could have developed the same idea for purely mathematical reasons. (It must be noted that this is a report on a talk from almost thirty years ago, and hence might be more of a false memory on my part than an accurate reporting.)

21.5. Yang-Mills Equations

We now want to outline Yang-Mills theory.   Its influence on modern mathematics can hardly be overestimated. We will use the language of the Lagrangian Machine.

Let E be a vector bundle of rank k on an n-dimensional manifold M. Let ω be a connection matrix for E with respect to some basis of sections for E. There is the corresponding $k \times k$ curvature matrix Ω whose entries are 2-forms on the base manifold M. Suppose a metric exists on M. This will allow us to define a Hodge $\star$ operator, which as we saw earlier, will map 2-forms to $(n-2)$-forms. Then we can form a new $k \times k$ matrix consisting of $(n-2)$-forms, namely, $\star\Omega$.

We are now ready for the key definition.

Definition 21.5.1. *The Yang-Mills Lagrangian is the map*

$$\mathcal{YM} : Connections\ on\ E \rightarrow \mathbb{R}$$

defined by

$$\mathcal{YM}(\omega) = \int_M \mathrm{Trace}(\Omega \wedge \star\Omega).$$

A connection that is a critical point of this Lagrangian is called a Yang-Mills connection. The Euler-Lagrange equations corresponding to this Lagrangian are the Yang-Mills equations.

The manifolds and bundles that are studied are such that the Yang-Mills Lagrangians are well defined. Note at the least that $(\Omega \wedge \star\Omega)$ is a matrix of n-forms and hence the trace is a single n-form on M, which can indeed be integrated out to get a number.

---

[1] I suspect that he was actually thinking of buildings at SUNY at Stony Brook. In the early 1970s, Yang from the Stony Brook Physics Department did start to talk to James Simon of the Stony Brook Math Department.

The connections that we study are those are the critical points of $\mathcal{YM}$. The corresponding Euler-Lagrange equations are the *Yang-Mills differential equations*.

If we take physics seriously, then we should expect that Yang-Mills connections should be important in mathematics, even in non-physical situations. This indeed happens. In particular, Donaldson [14, 15, 16, 36], concentrating on four-dimensional simply connected manifolds with $SU(2, \mathbb{C})$ vector bundles, discovered beautiful mathematical structure that no one had previously suspected.

This process has continued, such as in the equally groundbreaking work of Seiberg-Witten theory [60, 45, 64, 48, 42, 39]. And this revolution will continue.

Bibliography

[1] M. Atiyah, "On the Work of Simon Donaldson," *Proceedings of the International Congress of Mathematicians*, Berkeley CA, 1986, American Mathematical Society, pp. 3–6.

[2] M. Atiyah, "On the Work of Edward Witten," *Proceedings of the International Congress of Mathematicians*, Kyoto, 1990 (Tokyo, 1991), pp. 31–35.

[3] M. Atiyah, *Collected Works* Volume 6, Oxford University Press, 2005.

[4] Stephen J. Blundell, *Magnetism: A Very Short Introduction*, Oxford University Press, 2012.

[5] R. Bott, *Collected Papers of Raoul Bott*, Volume 4, edited by R. MacPherson, Birkhäuser, 1994.

[6] William Boyce and Richard DiPrima, *Elementary Differential Equations and Boundary Value Problems*, eighth edition, Wiley, 2004.

[7] J. Buchwald, Electrodynamics from Thomson and Maxwell to Hertz, Chapter 19 in *The Oxford Handbook of The History of Physics* (edited by J. Buchwald and R. Fox), Oxford University Press, 2013.

[8] J. Buchwald and R. Fox (editors), *The Oxford Handbook of the History of Physics*, Oxford University Press, 2013.

[9] George Cain and Gunter Mayer, *Separation of Variables for Partial Differential Equations: An Eigenfunction Approach* (Studies in Advanced Mathematics), Chapman & Hall/CRC, 2005.

[10] S. S. Chern, W. H. Chen and K. S. Lam, *Lectures in Differential Geometry*, World Scientific, 1999.

[11] J. Coopersmith, *Energy, the Subtle Concept: The Discovery of Feynman's Blocks from Leibniz to Einstein*, Oxford University Press, 2010.

[12] H. Corben and P. Stehle, *Classical Mechanics*, second edition, Dover, 1994.

[13] O. Darrigol, *Electrodynamics from Ampère to Einstein*, Oxford University Press, 2000.

[14] S. Donaldson, "An application of gauge theory to 4-dimensional topology," *Journal of Differential Topology*, Volume 18, Number 2 (1983), pp. 279–315.

[15] S. Donaldson, "Connections, cohomology and the intersection forms of 4-manifolds," *Journal of Differential Geometry*, Volume 24, Number 3 (1986), pp. 275–341.

[16] S. Donaldson and P. Kronheimer, *The Geometry of Four-Manifolds*, Oxford University Press, 1990.

[17] B. d'Espagnat, *Conceptual Foundations of Quantum Mechanics*, second edition, Westview Press, 1999.

[18] P. Dirac, *Principles of Quantum Mechanics*, fourth edition, Oxford University Press, 1958 (reprinted 1981).

[19] A. Einstein et al., *The Principle of Relativity*, Dover, 1952.

[20] Lawrence Evans, *Partial Differential Equations*, Graduate Studies in Mathematics, Volume 19, American Mathematical Society, 1998.

[21] R. Feynman, R. Leighton and M. Sands, *The Feynman Lectures on Physics* Volume 1, Addison-Wesley, 1963.

[22] G. Folland, *Introduction to Partial Differential Equations*, Mathematical Notes, Vol. 17, Princeton University Press, 1976.

[23] G. Folland, *Quantum Field Theory: A Tourist Guide for Mathematicians*, Mathematical Surveys and Monographs, Volume 149, American Mathematical Society, 2008.

[24] G. Folland, *Real Analysis: Modern Techniques and Their Applications*, Wiley, 1999.

[25] A. P. French, *Special Relativity*, Chapman & Hall, 1989.

[26] T. Garrity, *All the Mathematics You Missed but Need to Know for Graduate School*, Cambridge, 2002.

[27] J. Gray, *Henri Poincaré: A Scientific Biography*, Princeton University Press, 2012.

[28] B. Greene, *The Elegant Universe*, Vintage Books, 2000.

[29] P. Gross and P. R. Kotiuga, *Electromagnetic Theory and Computation: A Topological Approach*, Mathematical Science Research Institute Publication, 48, Cambridge Unversity Press, 2004.

[30] V. Guillemin and A. Pollack, *Differential Topology*, American Mathematical Society, reprint edition, 2010.

[31] D. M. Ha, *Functional Analysis.* Volume I: *A Gentle Introduction*, Matrix Editions, 2006.

[32] D. Halliday and R. Resnick, *Physics*, third edition, John Wiley and Sons, 1977.

[33] J. H. Hubbard and B. B. Hubbard, *Vector Calculus, Linear Algebra, and Differential Forms: A Unified Approach*, Prentice Hall, 1999.

[34] F. Jones, *Lebesgue Integration on Euclidean Space*, Jones and Bartlett Learning; revised edition, 2000.

[35] Y. Kosmann-Schwarzbach, *The Noether Theorems: Invariance and Conservation Laws in the Twentieth Century*, Springer-Verlag, 2011.

[36] H. B. Lawson Jr., *The Theory of Gauge Fields in Four Dimensions*, American Mathematical Society, 1985.

[37] P. Lorrain and D. Corson, *Electromagnetic Fields and Waves*, second edition, W. H. Freeman and Company, 1970.

[38] G. Mackey, *Mathematical Foundations of Quantum Mechanics*, Dover, 2004.

[39] M. Marcolli, *Seiberg-Witten Gauge Theory*, Hindustan Book Agency, 1999.

[40] J. McCleary, "A topologist's account of Yang-Mills theory," *Expositiones Mathematicae*, Volume 10 (1992), pp. 311–352.

[41] P. Milonni, *The Quantum Vacuum: An Introduction to Quantum Electrodynamics*, Academic Press, 1994.

[42] J. Moore, *Lectures on Seiberg-Witten Invariants*, Lecture Notes in Mathematics, Volume 1629, Springer-Verlag, 2001.

[43] T. Moore, *A Traveler's Guide to Spacetime: An Introduction to the Special Theory of Relativity*, McGraw-Hill, 1995.

[44] F. Morgan, *Real Analysis and Applications: Including Fourier Series and the Calculus of Variations*, American Mathematical Society, 2005.

[45] J. Morgan, *The Seiberg-Witten Equations and Applications to the Topology of Smooth Four-Manifolds*, Princeton University Press, 1995

[46] A. Moyer, *Joseph Henry: The Rise of an American Scientist*, Smithsonian Institution Press, 1997.

[47] D. Neuenschwander, *Emmy Noether's Wonderful Theorem*, Johns Hopkins University Press, 2011.

[48] L. Nicolaescu, *Notes on Seiberg-Witten Theory*, Graduate Studies in Mathematics, Vol. 28, American Mathematical Society, 2000.

[49] P. Olver, *Applications of Lie Groups to Differential Equations*, second edition, Graduate Text in Mathematics, Vol. 107, Springer-Verlag, 2000.

[50] C. O'Raifeartaigh (editor), *The Dawning of Gauge Theory*, Princeton University Press, 1997.

[51] A. Pais, *Subtle Is the Lord: The Science and the Life of Albert Einstein*, Oxford University Press, 1982.

[52] A. Pais, *Niels Bohr's Times: In Physics, Philosophy, and Polity*, Oxford University Press, 1991.

[53] A. Pais, *Inward Bound*, Oxford University Press, 1988.

[54] H. Poincaré, "The Present and the Future of Mathematical Physics," *Bulletin of the American Mathematical Society*, 2000, Volume 37, Number 1, pp. 25–38.

[55] J. Powell and B. Crasemann, *Quantum Mechanics*, Addison-Wesley, 1961.

[56] H. L. Royden, *Real Analysis*, Prentice Hall, 1988.

[57] W. Rudin, *Real and Complex Analysis*, McGraw-Hill Science/Engineering/Math, third edition, 1986.

[58] W. Rudin, *Functional Analysis*, McGraw-Hill Science/Engineering/Math, second edition, 1991.

[59] G. Simmons, *Differential Equations with Applications and Historical Notes*, McGraw-Hill, 1972.

[60] N. Seiberg and E. Witten, "Monopole Condensation and Confinement in $N = 2$ Supersymmetric Yang-Mills Theory," *Nuclear Physics*, B426, (1994).

[61] M. Spivak, *Calculus on Manifolds: A Modern Approach to Classical Theorems of Advanced Calculus*, Westview Press, 1971.

[62] F. Steinle, Electromagnetism and Field Theory, Chapter 18 in *The Oxford Handbook of The History of Physics* (edited by J. Buchwald and R. Fox), Oxford University Press, 2013.

[63] S. Sternberg, *Group Theory and Physics*, Cambridge University Press, 1994.

[64] C. Taubes and R. Wentworth, *Seiberg-Witten and Gromov Invariants for Symplectic 4-Manifolds*, International Press of Boston, 2010.

[65] R. Tolman, *Relativity, Thermodynamics and Cosmology*, Oxford University Press, 1934.

[66] R. A. R. Tricker, *The Contributions of Faraday and Maxwell to Electrical Science*, Pergamon Press, 1966.

[67] F. Verhulst, *Henri Poincaré: Impatient Genius*, Springer-Verlag, 2012.

[68] R. O. Wells Jr., *Differential Analysis on Complex Manifolds*, third edition, Springer Verlag, 2010.

[69] T. Wu and C. Yang, "Concept of nonintegrable phase factors and gobal formulation of gauge fields," *Physical Review D*, Volume 12, Number 12 (1975), pp. 3845–3857.

[70] E. Zeidler, *Quantum Field Theory. I: Basics in Mathematics and Physics*, Springer-Verlag, 2006.

[71] E. Zeidler, *Quantum Field Theory. II: Quantum Electrodynamics*, Springer-Verlag, 2009.

[72] E. Zeidler, *Quantum Field Theory. III: Gauge Theory*, Springer-Verlag, 2009.

Index

acceleration, 46, 48, 57, 61, 70, 79, 83
  under Lorentz transformations, 45
action, 71, 81
adjoint, 150–152, 162, 181, 185, 195
alternating forms, 248, 249
amplitude, 3, 163, 164, 178, 193
annihilation operator, 179, 181, 185, 196
Atiyah, Michael, 6

basis, 124, 160, 218, 229, 230, 234–236, 241,
    242, 244, 247–253, 255, 256, 261, 262
  for $\Lambda^k(\mathbb{R}^n)$, 104, 106, 120, 122–126, 128,
    129, 131, 139
  Hamel, 160
  Hilbert space, 147–149, 161
  orthonormal, 147–149, 161, 168
  Schauder, 147, 148, 160
Blundell, Stephen J., 7
Born, Max, 3
Bott, Raoul, 6, 8
Boyce, William, 190
Buchwald, Jed, 7
bundle
  fiber, 215, 217
  principal, 219
  tangent, 222–224, 229
  trivial, 238
  vector, 91, 214, 216, 217, 219, 220, 222,
    232–235, 238, 239, 241, 243, 244, 246,
    253–255, 257–259, 261, 262, 267–273

Cain, George, 190
calculus of variations, 70, 71, 77, 78, 80, 130,
    132, 133
Cauchy sequence, 143–146, 157, 158
Cauchy-Schwarz inequality, 167

chain rule, 23, 73, 74, 84, 103, 137, 225, 229
charge invariance, 63
commutator, 156, 171
complete inner product space, 143, 144, 157,
    159, 175
connection, 5, 91, 233–240, 242, 243, 254,
    257–259, 261, 262, 266–271
  for Maxwell's equations, 263–265, 267
  and curvature, 258
  trivial, 239, 254, 261
  Yang-Mills, 5, 6, 266, 272
conservation of energy, 83, 85
conservative vector field, 67, 68, 79
Coopersmith, Jennifer, 46
coordinate charts, 206
Corson, Dale, 62
Coulomb gauge, 187, 189
Coulomb's law, 1, 56, 59, 62, 63, 65
covariant derivative, 240, 243
creation operator, 179, 181, 185, 196
critical curves, 72
curl, 10–15, 91, 119, 121
current, 2, 9, 12, 17, 20, 21, 25, 62, 134, 186,
    187, 189, 263, 269
current 1-form, 130, 131, 264, 265
curvature, 5, 201, 214, 257, 258, 261, 262,
    264–267, 269–271
  from connections, 258
  definition, 258
  in beginning calculus, 257
  matrix, 259
  Maxwell's equations, 263–265, 267

density, 17
  charge, 9, 263, 265
  energy, 193

density (*cont.*)
  magnetic, 11
differential forms, 103, 107, 109, 115, 119,
    120, 130, 131, 250, 264, 265
  and vector fields, 121
  as tensors, 248
  elementary, 104
  integrating k-forms, 114
DiPrima, Richard, 190
Dirac, P. A. M., 3, 171
divergence, 10, 13–15, 119, 121
Divergence Theorem, 10
Donaldson, Simon, 6, 273

eigenvalue, 150
eigenvector, 166
Einstein, Albert, 2, 7, 27, 33, 164,
    186
electromagnetic
  force, 58, 98, 99, 265
  2-form, 130, 131, 139, 264
  waves, 2, 3, 17, 20–22, 28, 163, 164, 186,
    187, 193, 195
energy, 3, 46, 47, 80, 84, 85, 87, 163, 164,
    170, 178, 182, 183, 193, 196
  conservation of, 83, 85
  density, 193
  energy-momentum, 48
  kinetic, 80, 82, 83, 164, 178, 179
  potential, 79, 80, 82, 83, 87, 164, 170, 178,
    179
Euler-Lagrange equation, 71, 73, 75–77, 82,
    83, 99, 101, 102, 134, 136, 139–141, 264,
    272
Evans, Lawrence, 190
exterior algebra, 103, 119, 120, 124
exterior derivative, 106, 121
extreme values, 72

Faraday 2-form, 130
Faraday, Michael, 2
Feynman, Richard, 83
fiber bundle, 215, 271
Folland, Gerald, 189
force, 56, 57, 62, 64, 69, 70, 79, 83, 165, 178,
    267, 269
  conservative, 67, 79
  electromagnetic, 4, 56, 58, 59, 98, 99, 265,
    269, 270
  gravitational, 4, 57, 58, 68, 80, 269
  special relativistic definition, 61, 62

  strong, 4, 71, 88, 133, 267, 269, 270
  under Lorentz transformations, 61
  weak, 4, 71, 88, 133, 267, 269, 270
Fourier series, 149
framing, 218, 234, 236
French, A. P., 32

Galilean transformations, 31
Galvani, Luigi, 1
Gilbert, William, 1
Glashow, Sheldon, 4
gradient, 13, 79, 83, 119, 121, 205
Gram-Schmidt process, 148, 161
Gray, Jeremy, 8
Greene, Brian, 8
Gross, Paul, 139
group, 53, 85
  action, 268
  automorphism, 219
  general linear, 216
  Lie, 219
  orthogonal, 32
  permutations, 124
  unitary, 268, 270
Guillemin, Victor, 89

Halliday, David, 56
Hamiltonian, 170, 178, 179, 181, 182, 185,
    186, 194, 196
harmonic function, 189
harmonic oscillator, 170, 176, 178, 179,
    185–187, 193–195
  energy of, 178, 182
Hausdorff, 207
Heisenberg, Werner, 3
Hermitian operator, 142, 149–152, 155, 156,
    162, 165–169, 172–174, 185,
    199, 200
  adjoint, 150
  eigenvalues of, 151, 166
Hertz, Heinrich Rudolf , 2, 20
Hilbert space, 142, 144, 147, 149–153,
    157–161, 165–168, 170–172, 174, 179,
    180, 184, 185, 199
  $L^2[0,1]$, 146, 149
  l_2, 145, 149, 159
  basis, 147–149, 161
Hodge $\star$ operator, 119, 127–129, 272
homogeneous polynomials, *see* symmetric
    tensors
Hubbard, Barbara, 115

Hubbard, John, 115
hydroelectric dam, 12

inner product, 122–126, 128, 129, 142–145,
    147, 151, 154, 157, 161, 162, 166–168,
    172, 175, 180, 185
  complete, 143, 144, 157, 159, 175
integration by parts, 73, 75, 93, 138
invariant, 37, 85

Jacobian, 111, 114, 118, 203, 204, 225, 230
Jordan, Pascual, 3

kinetic energy, 80, 178, 179
Kobayashi, Shoshichi, 6
Kosmann-Schwarzbach, Yvette, 85
Kotiuga, P. Robert, 139

Lagrangian, 70, 71, 78, 80–83, 85, 87, 88,
    98–102, 130, 132, 136, 176, 179, 265,
    267–269, 272
  electromagnetic, 98, 99, 133, 136, 140, 264
  inverse problem for, 101
Laplacian, 18
Lebesgue measure, 159, 170
Leibniz's rule, 227, 233, 234
light cone, 40
light-like, 40
linear
  change of coordinates, 34, 251
  differential equation, 25
  operator, 149, 151, 174, 175, 180, 199
Lorentz contractions, 35, 36
Lorentz metric, 38
Lorentz transformations, 31, 32, 35, 37–39,
    41, 43, 45, 52, 53, 62, 63, 68
  acceleration, 45
  force, 61
  mass, 48
  velocity, 43
Lorentz, Hendrik, 8
Lorrain, Paul, 62
lowering operator, 181

Möbius Strip, 220, 222
magnetism, 63
manifold, 130, 201, 202, 206, 211–223,
    232–235, 238–240, 244, 246, 253, 254,
    257, 258, 261, 262, 265–268, 272, 273
  abstract, 201, 206, 212, 213, 224, 227–229
  four-dimensional, 5

functions on, 212
  implicit, 201, 205, 212
  parametric, 113, 118, 201, 203, 212, 213,
    223–225
Marathe, Kishore, 6
mass
  relativistic, 48
  rest, 48
Maxwell's equations, 3, 5, 6, 9–11, 15, 20–22,
    25, 27, 28, 56, 58, 62, 70, 88, 89, 91, 96,
    97, 103, 119, 131, 132, 134, 136, 139,
    142, 186, 189, 195, 201, 214, 238, 239,
    263, 267, 268
  vacuum, 20
  nabla form, 14
  via differential forms, 130, 131, 264, 265
Maxwell, James Clerk, 1–3, 20
Mayer, Gunter, 190
Mills, Robert, 4
Milonni, Peter, 179
Minkowski length, 39
Minkowski metric, 38, 39, 119, 125, 129, 131,
    139
Minkowski, Hermann, 8
momentum, relativistic, 46
monochromatic, 186, 194, 195
Moore, Thomas, 46

nabla, 12
Neuenschwander, Dwight, 85
Newton, Isaac, 56, 57, 81
Noether's Theorem, 85

O'Raifeartaigh, Lochlainn, 8
Oersted, Hans Christian, 2
operator
  linear, 149, 151, 174, 175, 180, 199
  $q(f)$ and $p(f)$, 155
  spectrum, 150, 152, 166

Pais, Abraham, 7, 164
parallel transport, 243
path integral, 79, 80, 110, 118
permutation, 105, 124, 250
photoelectric effect, 3, 142, 163, 164, 186, 196
Planck constant, 164, 170
Poincaré's Lemma, 107
Poincaré, Henri, 8
Pollack, Alan, 89
potential, 16, 80, 88, 91, 131–133, 267
  Coulomb, 187, 189

potential (*cont.*)
  energy, 79, 80, 83, 87, 164, 170, 178, 179
  1-form, 131, 132, 136, 140, 264
  scalar, 88, 89, 91, 97–100, 102, 187–189
  vector, 88, 89, 91, 97–100, 102, 187, 188, 196
  proper time, 36, 47, 54

quantization, 163, 170, 176, 186, 195, 271

raising operator, 181
relativistic displacement, 38, 39, 46, 47
relativistic mass, 48
relativistic momentum, 46, 54
Resnick, Robert, 56

Salem, Abdus, 4
Schrödinger's equation, 169, 170
Schrödinger, Erwin, 3
Schwartz space, 153, 154, 157, 158, 160, 172, 185
section, 218, 232–235, 238, 239, 241, 243, 244, 246, 253, 254, 258–262, 267, 268, 272
  covariant derivative of, 240, 243
separation of variables, 190
Simon, James, 272
space-like, 39–42
special relativity, 8, 21, 23, 25, 27, 33, 39, 59–63, 85, 125
spectrum, 150, 152, 166
springs, 170, 176, 178, 186
square integrable functions, 146, 170
Steinle, Fredrich, 7
Sternberg, Shlomo, 267
Stokes Theorem, 10, 11
symmetric tensors, 250, 251

tangent, 11, 67, 103, 109, 224, 229, 240, 241, 243, 244, 257
  bundle, 222–224, 229
  line, 224, 225
  plane, 111, 223–226, 230
  space, 114, 202, 224, 225, 227, 230
Taubes, Clifford, 6

tensor product, 233, 247
  alternating forms, 248, 249
  linearizations of bilinear maps, 251
  symmetric tensors, 250, 251
time dilation, 35, 36
time-like, 39, 40, 42
transformations
  Galilean, 31
  Lorentz, 31, 32, 35, 37–39, 41, 43, 45, 52, 53, 61–63, 68
transition functions, 217, 238
translations, *see* transformations

Uhlenbeck, Karen, 6

vacuum, 9, 20, 27, 179, 183
vector bundle, 91, 214, 216, 217, 219, 220, 222, 232–235, 238, 239, 241, 243, 244, 246, 253–255, 257–259, 261, 262, 267–273
velocity, 32, 35, 38, 40, 46, 48, 49, 51, 57–61, 68, 78, 80, 100, 178
  light, 30
  under Lorentz transformations, 43
Verhulst, Ferdinand, 8
Volta, Alessandro, 1
von Neumann, John, 3

wave
  amplitude, 3, 163, 164, 178, 193
  electromagnetic, 2, 3, 17, 20–22, 28, 163, 164, 186, 187, 193, 195
  equation, 17–19, 21, 25, 26, 189, 198
  equation for vector fields, 19
Weinberg, Steven, 4
Weyl, Herman, 4, 5, 8
Witten, Edward, 7, 8
Wu, Tai Tsun, 5, 270

Yang, Chen-Ning Franklin, 4, 5, 270–272
Yang-Mills, 5, 6, 266, 267, 271, 272
Yau, Shing-Tung, 6

Zeidler, Eberhard, 6, 85

Printed in the United States
By Bookmasters